本色

顶级别墅空间 Top Villa Space

TRUE COLOURS

深圳市博远空间文化发展有限公司 编

華中科技大學出版社
http://www.hustp.com

Preface

Villa, what is it? The explanation is the dictionary, villas, i.e. don't industry, outside the habitat home dwelling place, is a specific type of habitation. It is a way of life, a kind of extreme way of life, is a big endowment capital to childhood to see the top of the way of life. And the example room, it is the real estate enterprise to showcase its product platform, a good example that stimulates the potential customers to purchase desire, enhance the brand image building itself, and then drive around with value related industries, achieve closure marketing strategy. Villa example is the perfect union, it shows us the contemporary villa design the newest idea and the latest trends. Villa example room can be low-key costly, big hidden in city: also can be DongLi adopt chrysanthemum, carefree hanshan. Life - not expensive. Quiet and tastefully laid out - not expensive. Taste -- not expensive. True villas, necessarily mean low density, means quality, means love nature, enjoy life, means that environmental protection and oneness.

Villa example room design bearing lead mass live trend function, coupled with the media and public attention, developers enough budget investment, these all let designers have more display space, to a certain extent, gave stylist wider realize the conceptual design of space, in unfettered cases and better will have specific design ideas. In recent years, along with the continuous development of real estate in China with mature, example room almost become a specially designed the fields, some focus on example room domains designers also so radically, become much-talked-about star characters. And, more than that of favor. Villa example room This book will invite to countries well-known designers, make the latest example villa most authoritative villa design, lead villa design example new trend. Books from China (including party included mainland, Taiwan and Hong Kong) and foreign dozens of designer's latest example room, more than 100 sets of works covered romantic style, full of full-bodied breath French style of American style, rich flavor of Chinese style style, typical Mediterranean style, languid is lazy relaxation of Thai style. Delicate British style, full-bodied amorous feelings of Spanish style, Bali style; The district distribution in Hong Kong and Taiwan, coastal and inland country, abroad; Design les genres, rich but not multifarious.

Reality always complicated. We are going through an increasingly diversified era, this time not only luxury and plain write, heavy and complicated and contracted coexistence, also has the most extraordinary splendour colorful life style, the most tumultuous blurred interpersonal relationship. Everything is a dazzling understand disorderly. A good books believe that will let you in dizzying life find comfort, find the mind the pleasures of life. Book might be a good choice for you. In this book, you can see pure European concept in the city's peak, also can enjoy looking Chinese space feelings have expressed the relocation claims sit. Walking in the book and find a set of their favorite villa! A good example of constituencies that published, design with pictures and documented the formation process of example room, and eternal frames example entity forms. It will make stylist art and life regeneration and the experimental results continuously sublimation, let its limited life to infinite continue. Believe that book has reached the expectations. Court meaning, nobles palace can deeply impressed, plain east, aesthetic, give yourself some time to sit down and savor some!

本色 顶级别墅空间 Top Villa Space TRUE COLOURS

序言

别墅，是什么？辞典上给出的解释是，别墅，即别业，宅居之外的居所，是一种特定类型的居所，也体现着一种极致的生活方式。而样板房，它是房地产企业展示自身产品的平台，一套好的样板房可以激发出潜在客户的购买欲望，提升楼盘自身的品牌形象与价值，进而带动周边相关产业的发展，成功实现多方共赢的营销策略。别墅样板房是二者的完美结合，它向我们展示的是当代别墅设计的最新理念和最新趋向。别墅样板房可以是低调奢华，大隐于市；也可以是采菊东篱，悠然南山。生活——不一定昂贵；幽雅——不一定昂贵；品味——不一定昂贵。真正的别墅，一定意味着低密度，意味着品质，意味着热爱自然、享受生活，意味着环境保护和天人合一。

别墅样板房设计担负着引领大众居住潮流的责任，加之媒体和公众的高度关注，以及开发商充足的预算投入，这些都在一定程度上给予了设计师更广阔的实现概念设计的空间，在不受束缚的情况下更好地将具有针对性的设计想法表达出来。近年来，随着中国房地产行业的不断发展与成熟，样板房设计几乎成为一个专门的领域，一些专注于样板房领域的设计师也因此声名鹊起，成为备受关注的明星人物，而别墅样板房更甚。本书诚邀各国知名别墅样板房设计师，打造最新、最权威的别墅样板房设计，引领别墅设计新潮流。书中共收录来自国内及国外的几十位设计师的最新样板房作品百余套，风格涵盖浪漫的法式风格、休闲的美式风格、富有韵味的中式风格、充满自然气息的地中海风格、慵懒放松的泰式风格、精致的英式风格、热情奔放的西班牙风格、风情万种的巴厘岛风格；设计风格丰富而不繁杂。

我们正在经历一个越来越多元化的时代，这个时代不仅奢华与质朴并立，繁复和简约共存，还拥有最为异彩纷呈的生活方式，最为纷乱迷离的人际关系。一切都令人眼花缭乱。我们相信一本优秀的图书一定会让您在纷杂的生活中找到心灵的慰藉，找到生活的乐趣。本书或许会是您不错的选择。在本书中，既可以看到纯粹的西式风格在城市的巅峰眺望，也能领略到中式风格所表述的安居主张。徜徉于此书中，找一套自己喜欢的别墅吧！一本优秀样板房设计的结集出版，是以图片和文字记录样板房的形成过程，并永恒定格样板房的实体形态。它将使设计师艺术与生活的实践成果不断再生与升华，让其有限生命得以无限延续。相信本书已经实现了这个期望。庭院深深，贵胄皇庭，质朴东方，唯美印象，给自己一点时间，坐下来，细细品味一番吧！

CONTENTS 目录

东方华贵 ORIENTAL LUXURY

当代美式 CONTEMPORARY AMERICAN

本色

顶级别墅空间Top Villa Space

TRUE COLOURS

ORIENTAL LUXURY

东方华贵

明丽温馨的色调
奢华高雅的材质
开阔大气的布局
高调地彰显着雍容贵气的身份

东方华贵
高雅奢侈
感观天生领袖派的气质
品味王者风范的高贵
引领从容而高雅的生活潮流

Tbdc Casa Demo

卡萨样板房

项目名称：卡萨样板房 / 设计师：黄鹏霖、黄怀德 / 设计单位：台北基础设计中心 / 空间面积：247.8 m^2
主要建材：复古面石材、银狐石、养锈铁件、纯白亚克力、镀钛钢板 / 项目地址：中国台湾台北 / 摄影师：王基守

Project Name: Tbdc Casa Demo / Designers: Huang Penglin, Huang Huaide / Design Company: Taipei Base Design Center
Area: about 247.8 m^2 / Materials: Stone With Retro Covering, Silver Fox Stone, Rustproof Iron, Pure White Acryl, Titanizing Steel Plate
Location: Taipei, Taiwan, China / Photographer: Black Wong

▶
此案为两层楼空间，利用挑空的手法，用锈蚀铁件延伸衔接两层楼，不仅创造出层次丰富的空间，视觉上也更为宽敞。为了使得空间更有弹性，区域分隔的部分，特别以推拉门以及玻璃取代传统隔断墙。有别于以往的电视墙，此案以不锈钢柱体来取代常见的电视墙，线槽及维修孔亦隐藏其中。
机能及收纳的每一个细节皆是经过精密计算及多重考量，为此空间量身定制。值得一提的是部分家具，都是由 TBDC 规划定制设计，包括更衣室衣柜、餐桌、多功能茶几、卫生纸置物架以及厨具是采用纯手工定制发色不锈钢柜体搭配特别的银狐石台面。区区两百多字或许无法完整描述此案的精粹，以及施工过程中的困难，但在不断克服、调整、取舍后，空间处处充满惊喜。

10 53

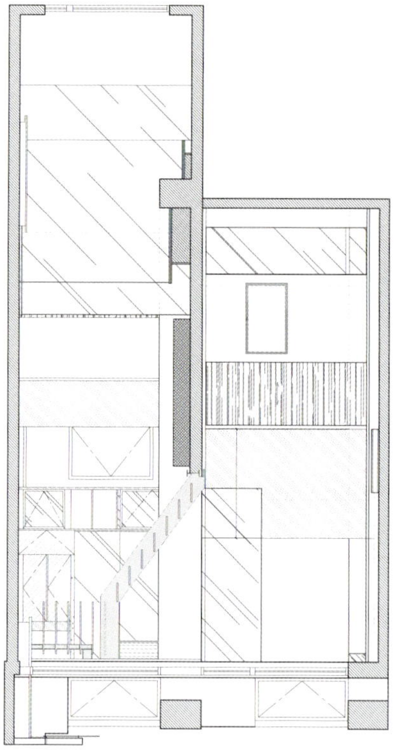

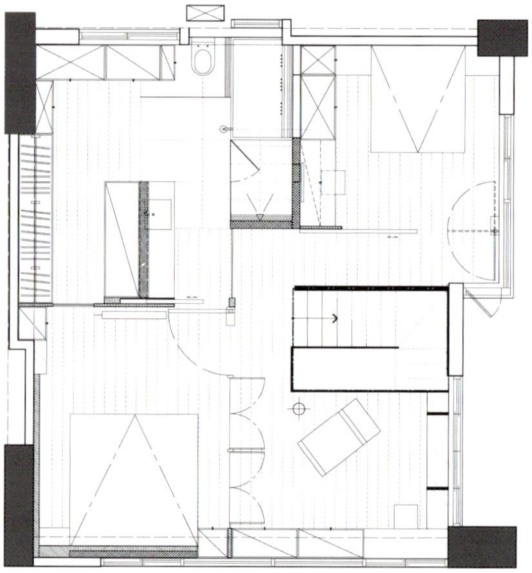

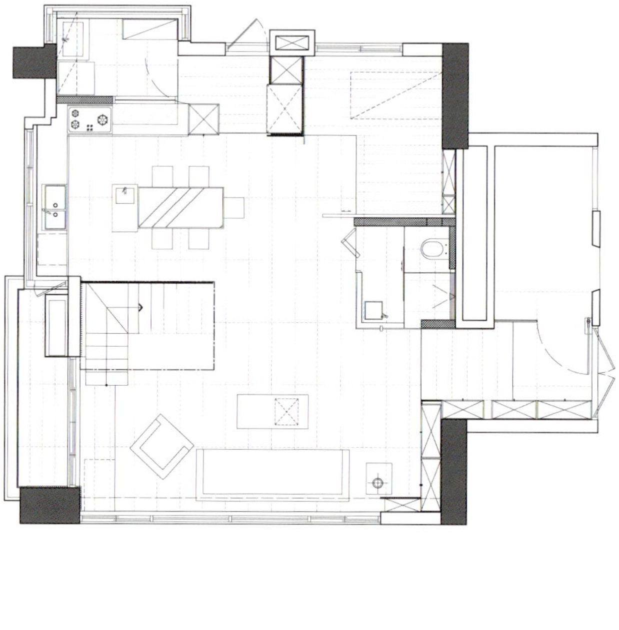

▶

This case is a 2-floor one. By using hollow-out structure, rustproof iron extends while connecting these 2 floors. It creates a sense of space with layers and presents to be a lot more spacious for vision. As for area separation, in order to make the space more flexible, sliding doors and glasses are used to substitute orthodox walls. What's different from the usual television wall is that in this case, architects use stainless steel cylinder to substitute frequently seen television wall, so that wire casing and maintenance finestra can be hidden within it.

Every detail of function and storage has been precisely calculated and thoughtfully considered, especially for this space. What's worth to be mentioned is some of the furniture, they are designed and made by TBDC, including wardrobe, dining table, multi-functional teapoy, toilet tissue holder and kitchen ware, which is wholly hand-made and with special silver fox table-board in stainless steel and hair color. Merely 200 words might not be able to describe the essence of this case, or complexity and difficulty during the construction process. However, architects constantly overcome, adjusted, selected, after that. Every aspect is full of surprise.

HOUSE LIKE A MUSEUM

像博物馆的住宅

项目名称　像博物馆的住宅 / 设计师　Edward Suzuki Associates / 设计单位　Edward Suzuki Associates Inc
空间面积　776.4 m^2 / 项目地址　日本神奈川县镰仓市

Project Name　House Like a Museum / Designer　Edward Suzuki Associates / Design Company　Edward Suzuki Associates Inc / Area　about 776.4 m^2 / Location　Kamakura City, Kanagawa Prefecture, Japan

这个由日本建筑师爱德华。铃木与大和房建合作建成的预制住房“艾迪屋”，它的主题是，“室内遇上室外”。房子总占地 776 m^2，其中间有一个直径约 15 m 的圆形天井。该设计的主旨是让每一处房间都与中央花园景观相邻。

The theme, “Go in to go out,” used in the prefabricated housing system “EDDI's House” designed and built in collaboration with Daiwa House Industry, was once again applied to the design of this house and, as a result, the rectangular silhouette of the house was pushed to the boundary limits of the 776 square meter property in the center of which was placed a 15-meter diameter circular patio. The thrust of the planning was to allow each and every room to face and to have a view of this central garden.

入口处的半独立的夹层阁楼是一个用来招待客人住宿的和式房间。客人房旁边，是主人的私人工作间。沿着入口往上是个倾斜的，连接着客厅、餐厅和厨房区域的走廊。延着楼梯往上便是主人卧室和私人卫生间。架高的区域有个通往二楼小孩房和家庭影院空间的楼梯。二楼卧室南面是一片宽大的、可以俯瞰下面天井的花园露台。

Near the entrance and up on a mezzanine in a semi-detached fashion is a Japanese Room able to accommodate occasional guests. Next to it and readily accessible from the entrance is the owner's Private Office. Leading up from the entrance is a sloping gallery-corridor connecting the Living Room, Fireplace, Dining, and Kitchen areas. Again, another ramp connects this one-room like space to the elevated Master Bedroom with its private water closet. From this elevated area a staircase leads to the Family/Home Theater space adjacent to the Children's Bedrooms and Family Bathroom on the second level. On the south of these bedrooms spread a spacious garden terrace overlooking the circular patio below.

▶

外墙材料基本上采用的是硅胶土，能够起到隔热作用。内墙和顶棚主要是用日本的灰泥粉饰，地面采用竹层夹板和石灰石材质。所有的内置或者移动家具也是由竹层夹板制成。用竹材代替木材越来越流行，也更加环保。同样，这个房子的设计也借用了日本传统住房设计智慧中的元素，比如用来区分室内和室外的“缘侧”（外围走廊）、“坪庭”（小小的中庭）、“新沐”（屋顶出挑）、“Tsuufuu”（交叉通风）；通过精心挑选常绿树木或者落叶树木，屋顶花园也可以起到隔热的作用——所有这些看似被动的设计有效地利用了自然界的能量，并且没有因为被迫的机械性方式去浪费能量。

Finish material for the exterior is basically diatomite over exterior insulation, while for the interior it is predominantly Japanese stucco-paint for walls and ceilings and either bamboo laminates or limestone for the floors. Bamboo laminates are used for all built-in and movable furniture as well. Bamboo, in contrast to wood, grows much more rapidly and hence is more eco-friendly. In the same vein, efforts have been made to “borrow” from and apply the wisdom of traditional Japanese house design vocabulary such as “Engawa” (peripheral corridor) that fuses or separates as required the inside and the outside, “Tsuboniwa” (tiny patios), “Hisashi” (roof overhangs), “Tsuufuu” (cross ventilation), selected use of evergreens and non-evergreens, roof gardens for extra geothermal insulation-all basically “passive” means of applying nature's energy efficiently instead of “forced” mechanical ways expending unnecessary energy.

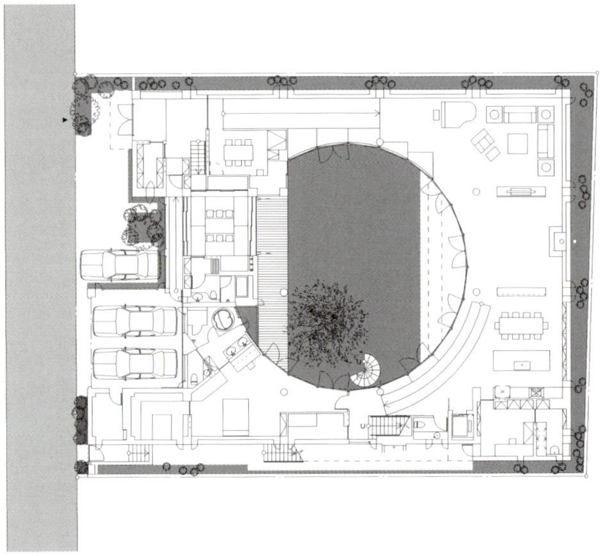

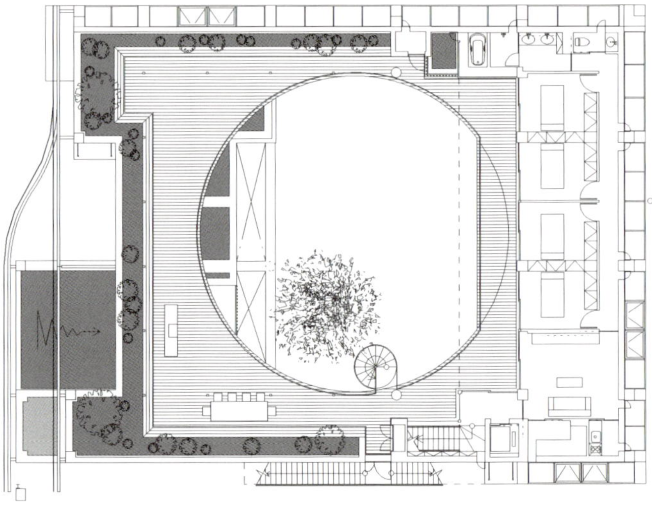

MANSOUREYA RESIDENCE

Mansoureya 别墅

项目名称 Mansoureya 别墅 / 设计单位 Karim Mekhtigian（炼金术设计工作室）
空间面积 480 m^2 / 项目地址 埃及开罗马迪 / 摄影师 Sherif Tamim

Project Name Mansoureya Residence / Design Company Karim Mekhtigian, Alchemy Design Studio
Area about 480 m^2 / Location Maadi, Cairo, Egypt / Photographer Sherif Tamim

业主想要改造和扩展原有的室内空间，以满足举办一些大型聚会等生活需求。为了在这个大客厅里可以进行多种活动，480 m^2 的空间被划分为用餐区、厨房和可以容纳一大群人娱乐的设备丰富齐全的休闲区。另外，在小角落的座位里可以享用下午茶，用木头和大理石材料以及极简的线条设计的壁炉，突出了整个空间设计的精心之处。

The owners chose to create an extension to their already-established home to express their lifestyle requirements and use it to host large gathering. Acting as a big living room with areas for different activities, the 480sqm features a dining area, a kitchenette and fully-equipped lounge with enough space to entertain a large group of people. In addition to the little seating corner perfect for afternoon tea, the fireplace highlights the layout with its linear design and overwhelming use of wood/marble.

整个空间的设计灵活，运用了自然材料和朴实色彩；特意选用的中性色调搭配花园的烂漫景致。进行空间设计时，设计师想尽可能借用宽阔的景域，因此设计了落地窗，从而使室内空间和室外自然美景进行完美互动。
室内空间和光线的运用，以及木材覆层和地板材质协调搭配打造了一种禅意、轻松的氛围。充满活力的淡淡的颜色突显了周围自然景致，同时也给休闲空间带来一丝生活的气息。
视觉的刺激与舒适的环境、艺术的元素相结合，让人无法忽视有着超大的定制吊灯的楼梯。不对称组合的木头和玻璃，以及大型的照明用具给这个现代设计增添了些许戏剧感。

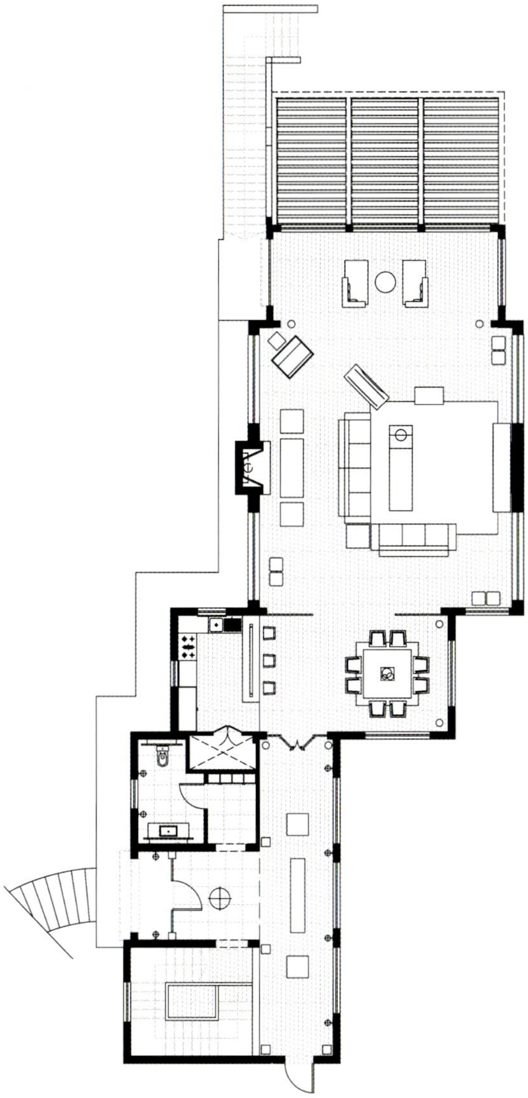

The whole volume is accentuated with a clever use of natural materials combined with an earthy color scheme; the neutral tones were intentionally selected to complement the garden view. While designing the space, Karim Mekhtigian wanted to make the most out of the expansive view, therefore insisting on merging the inside and outside through a window wall to overlook the natural interactive background.

The general mood of the room is a sum of its parts, the use of space and light in addition to the selection of wood cladding and flooring creates a certain synergy, which results in a Zen relaxing atmosphere. A touch of vibrant colors brings life to this casual living space as it borrows references from nature, once again emphasizing the surrounding view.

Combing comfort with visual exciting and artistic design elements, one cannot overlook the staircase with its oversized custom-made chandelier. Bringing together wood and glass in an asymmetrical composition, the large scaled lighting fixture adds a dramatic aspect to this modern design.

Lorenzo Residence

劳伦斯公馆

项目名称　劳伦斯公馆 / 设计师　Design Hirayama + Quesada / 设计单位　Design Hirayama + Quesada
空间面积　500 m² / 项目地址　菲律宾马尼拉 / 摄影师　Erik Liongoren

Project Name　Lorenzo Residence / Designers　Design Hirayama + Quesada / Design Company　Design Hirayama + Quesada
Area　about 500 m² / Location　Metro Manila, Philippines / Photographer　Erik Liongoren

劳伦斯公馆的业主是有两个孩子的年轻夫妇。本案采用的是现代亚洲设计风格，讲究的是质地自然的开放空间规划。虽然外形是现代和线性的，但是混合了例如菲律宾桃花心木、紫檀木和条纹乌木多种木质，从而使房间给人一种舒适而放松的感觉。为了使家更加生动有趣，整个空间采用了分层设计。顶棚采用的是凹凸结合设计。其他房间的色彩、窗台以及推拉门也都经过了精心的设计。

The Lorenzo Residence is home to a young couple with 2 children. It is done is an Asian contemporary style, playing on natural textures and an open space plan. While the shell is modern and linear, it is made cozy and relaxing by mixing different types of wood such as Philippine Mahogany, Narra, Kamagong. To make the space plan more interesting, layering was used. The ceiling uses a combination of coves and drops. The related spaces are carefully defined by color, window partitions and sliding doors.

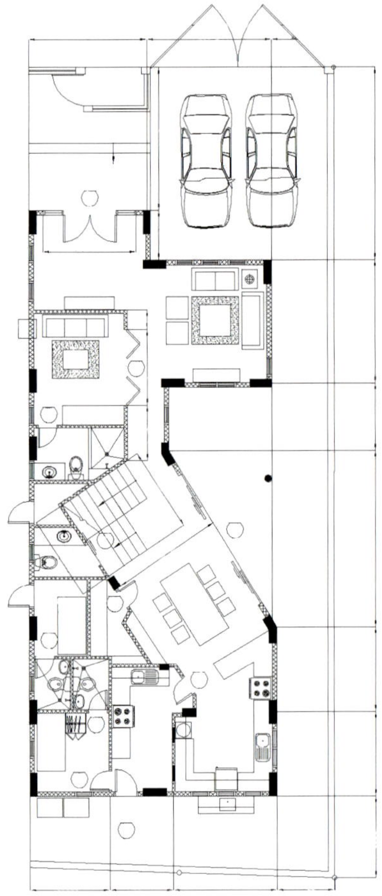

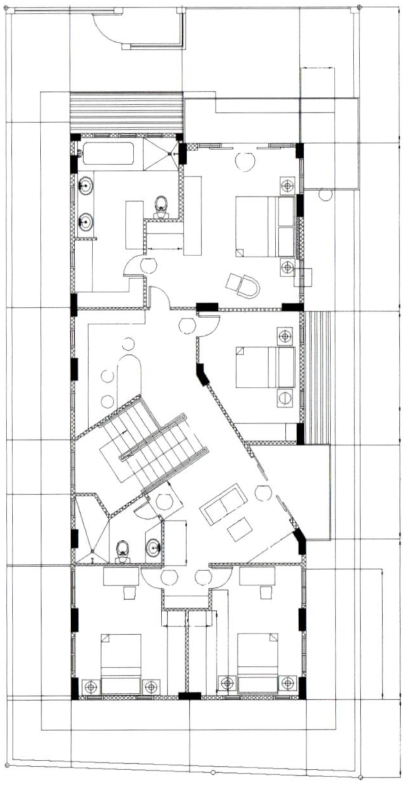

在这个规划中，通过一些特定的元素突出空间的层次感。为了显示这是一个年轻人的生活空间，在很多关键地方都用色彩来表现——开放式的厨房中的绿色和银灰色斑点，家庭休闲区的梅花和紫色，工作间的红色、绿色和蓝色等。楼梯直接设置在房间的中间，创造动感并且让自然光线可以照进来。房间的核心区是二楼的中庭和家庭休闲区，家庭休闲区是用来放松和娱乐的地方，整个空间给人一种现代、活泼，但又温馨、悠闲的感觉。

Each space can be seen in this open plan. There are some visual layers playing with the defining elements. To reflect the youth of the people living in the home, color is used in key areas - a splash of green and silver infused in the open kitchen, some plum and purple in the family, and primary red, blue and green in the study. The stairs are strategically placed prominently at the center of the house to create some movement and allow generous lights to come in. The heart of the home is the patio and the family room on the second floor. Here, the family gather to relax and play. The home reflects the family living in it – young and modern yet warm and casual.

TIANMU ZHANG RESIDENCE

天母张公馆

项目名称 天母张公馆 / 设计师 沈志忠 / [illegible]线设计 / 空间面积 198 m^2
主要建材 白色大理石、竹木皮、格子玻璃、实[illegible]不锈钢、竹地板、黑板岩 / 项目地址 中国台湾台北

Project Name Tianmu Zhang Residence / [illegible]en Chihchung (Ron Shen) / Design Company X-Line Design Company
Area about 198 m^2 / Materials White M[illegible], Leather, Lattice Glass, Solid Wood Shaft, Mirror Stainless Steel, Bamboo Flooring, Black Slate / Location Taipei, Taiwa[illegible]

▶
以“翻折”概念将各分属空间连接、转折与延伸，例如入口处到书房、两层楼的挑高壁面直至儿童房，皆以温润的木皮铺设；电视墙延伸到地板和入口处，以比较利落的白色大理石做串联；利用材质与动线的巧妙安排，模糊各空间的界限，不断开展新的视野及对话。

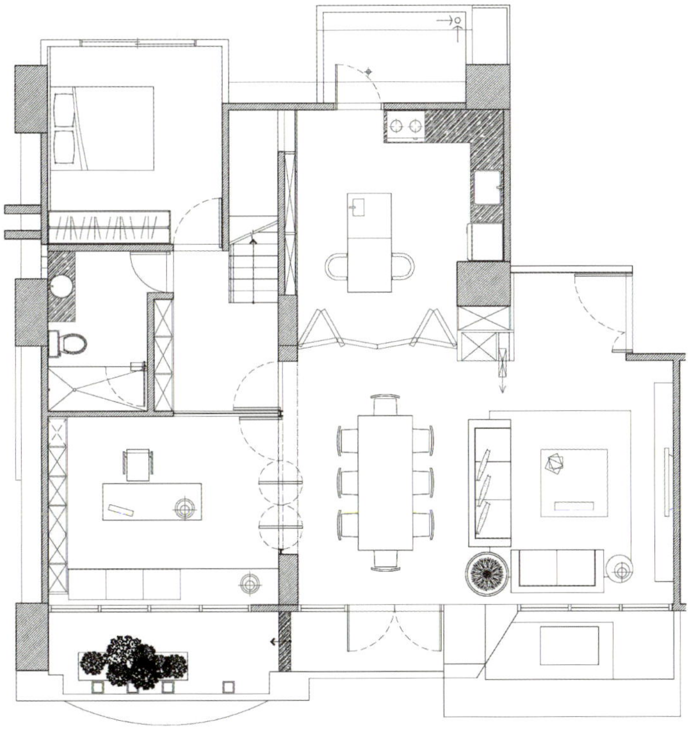

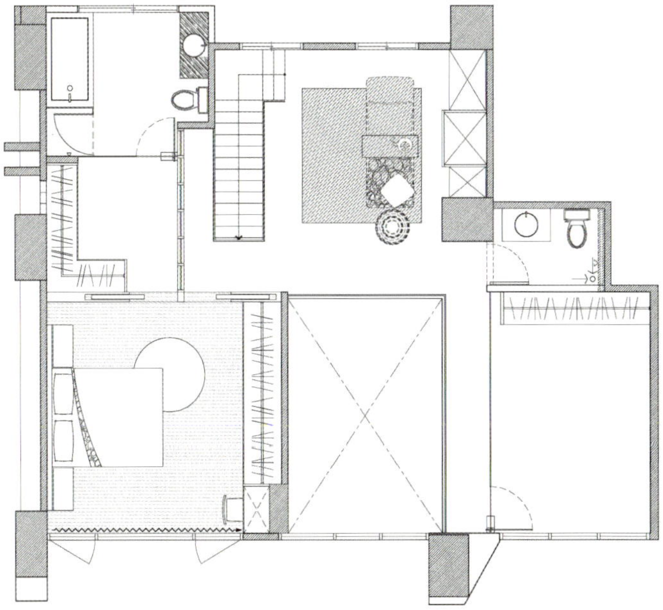

▶
The "turn-up" idea is adopted in the connection and extention of each area, for example, from the mezzanine to study room, both the wall between two floors and children's room are covered by the same wood; from the TV wall to floor and inner mezzanine are connected with cold brief marble....finely arrangement materials, vision connection and lines blurring the boundaries of spaces to develop the vision extension and creat space dialogue.

永和 YONGHE

项目名称 永和 / 设计师 黄千祝 / 空间面积 90 m² / 主要建材 进口壁纸、实木地板、铁件雷刻、特殊烤漆进口耐火板、玻璃、大理石、意大利砖 / 项目地址 中国台湾永和

Project Name Yonghe / Designer Huang Qianzhu / Area about 90 m² / Materials Imported Wallpaper, Hardwood Flooring, Metal Iron Mine Engraved, Special Imported Baked Paint Fireproof Veneer, Glass, Marble, Italian Tile / Location Yonghe, Taiwan, China

公共空间以凝聚家庭成员为设计主旨，打造一种时尚且奢华的空间氛围，独立却开放的书房工作区在客厅窗台边，客厅家具摆设以方便使用为主，整个空间设计以时尚奢华概念为主。

This case pursues the concept of modern luxury. The public area is the main design in the residence to enhance daily family connection. The study room is arranged by the windows of the living room as an open independent corner. Most furniture in the living room is movable. An exercising room meets the male house owner's needs for bodybuilding.

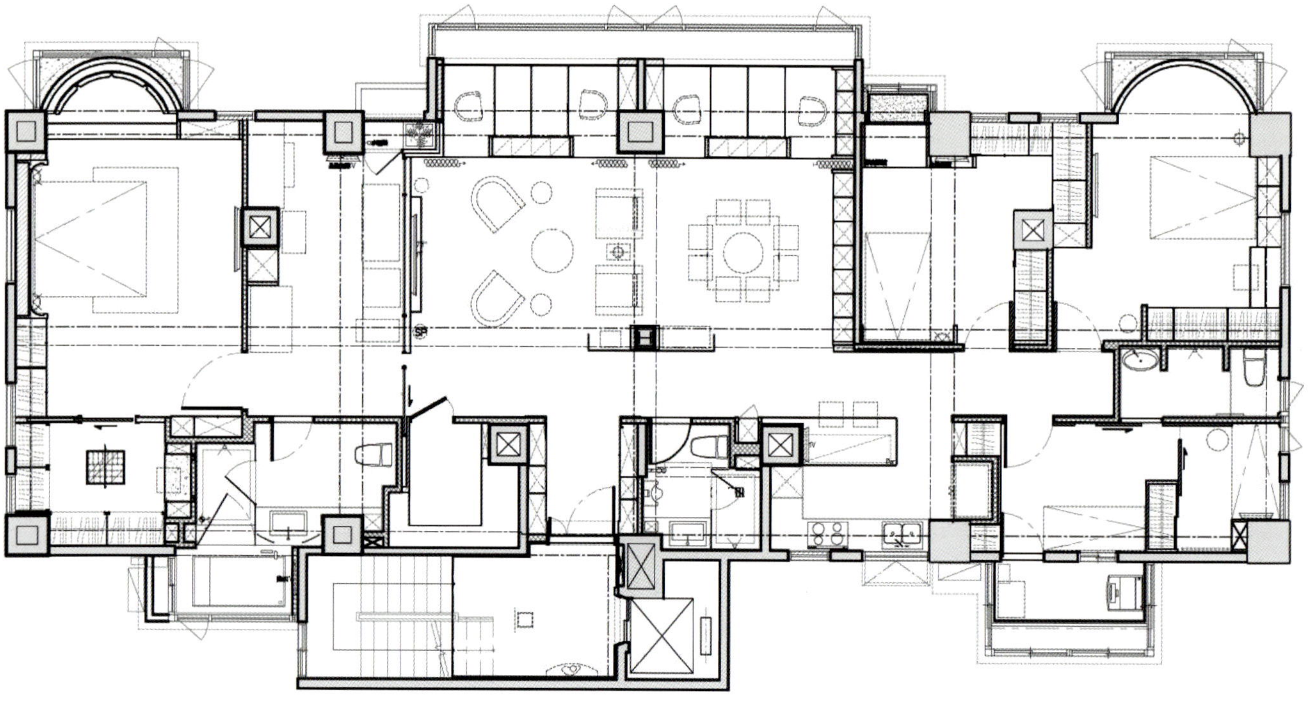

VERTIENTES HOUSE
斜坡上的房子

项目名称 斜坡上的房子 / 设计师 Pascal Arquitectos, Carlos Pascal, Gerard Pascal
设计单位 Pascal Arquitectos / 项目地址 墨西哥 Lomas de Chapultepec / 摄影师 Jaime Navarro

Project Name Vertientes House / Designers Pascal Arquitectos, Carlos Pascal, Gerard Pascal
Design Company Pascal Arquitectos / Location Lomas de Chapultepec México City / Photographer Jaime Navarro

▶

房屋有三层，占地 700 m²，主要空间如客厅、餐厅、厨房、书房和一个家庭房都布置在二层。一层设有停车室和服务区，三层有三个卧室，每个卧室配有卫生间和衣柜。各层之间有电梯相连，电梯可容纳10 个人。房子采用经典的设计概念，但是最后却综合了很多种设计风格。

使用的材料方面以经典风格为主，采用木地板、各式各样的大理石，如黄金天龙大理石、卡拉拉白色大理石和黑金花大理石。主门厅的黄金天龙大理石圆柱与其他家具整体的风格互相搭配，营造出精致而帅气的空间感。

照明方面采用较传统的方式，通过对茶几橱柜、台灯、烛台和普通灯泡的布置强调出艺术性，突出空间的氛围。

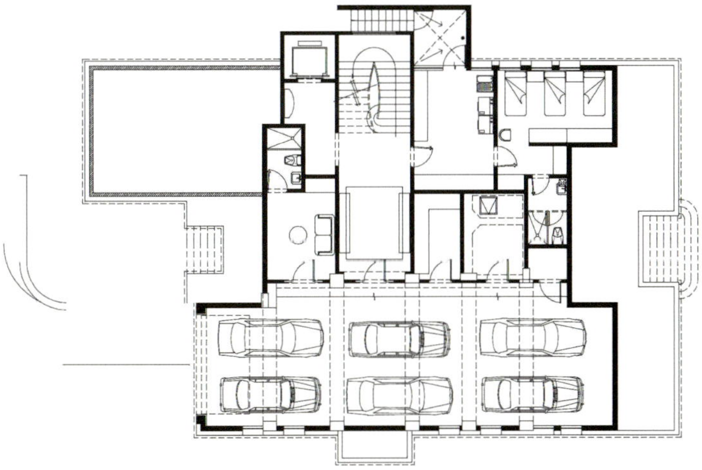

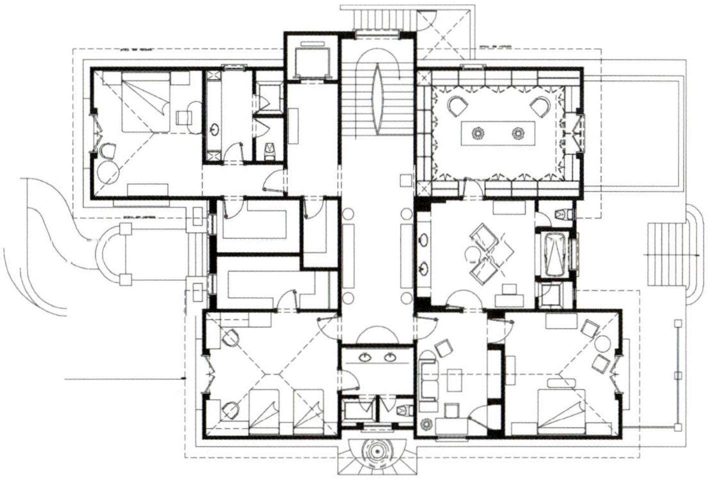

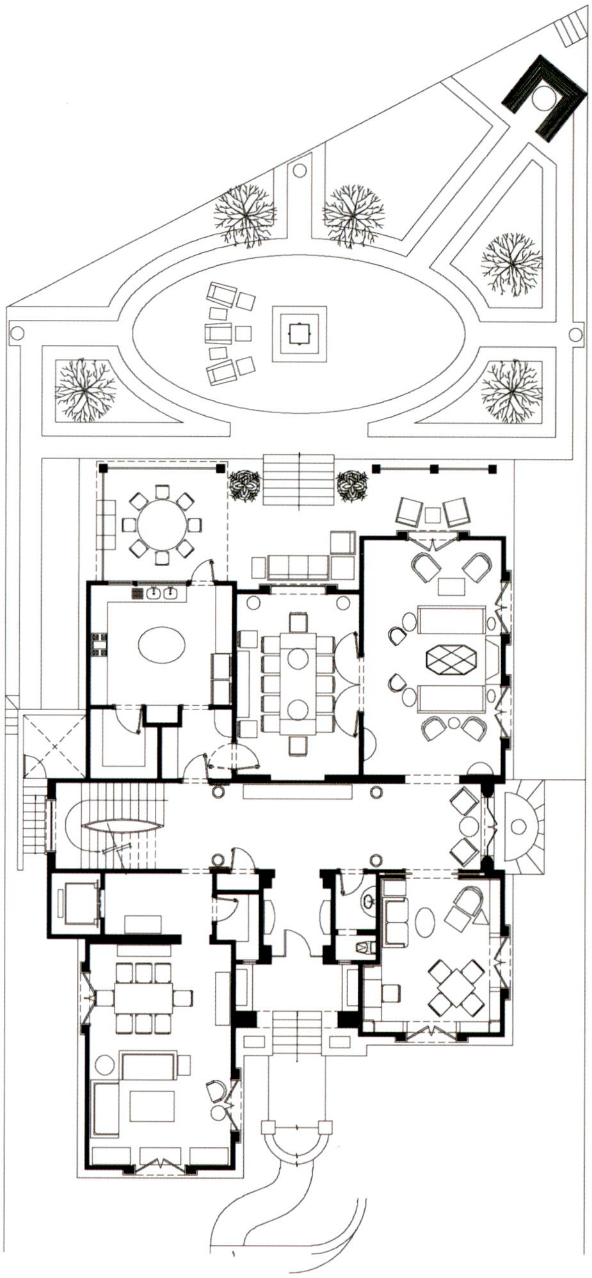

The house is composed of three levels, and occupies a total of 700sqm, the principal access, living room, dining room, kitchen, study and a family room belong to the middle level. The garage and the service zone belong to the lower level, and the higher level has three bedrooms, each with their bathroom and wardrobe. All levels communicate with an elevator for 10 people.This home was designed using classic concepts but interpreted like a synthesis of the same, styled and deformed to create a longevity effect, in the same way they are cool and contemporary.

The materials used refer to classic styles, wooden floors, different types of marble like Yellow "Triana", White "Carrara", and Black "Portoro", on the principal foyer solid columns of Yellow "Triana" marble, all conjugated with furniture, of style, they create total sophistication and coolness.

The illumination highlights the art and dramatizes its ambience, in which it implemented the use of lower technology, desk and foot lamps, sconce and common light bulbs.

TIANMU LI MANSION

▶ 天母李宅

项目名称　天母李宅 / 设计师　郭柏伸 / 设计机构　奇逸空间设计
空间面积　265 m² / 主要材料　大理石、实木格栅、玻璃砖、铁件、木地板

Project Name　Tianmu Li Mansion / Designer　Guo Boshen / Design Company　Free Interior Design
Area　about 265 m² / Materials　Marble, Wood Grill, Glass Bricks, Iron, Wood flooring

此案业主为年轻的夫妻以及两个小孩，由于原始格局、动线早已不敷使用，设计师在规划之初，即针对业主喜好、未来生活模式、基地状态等，重新从建筑体量细部至各室内空间作一完整而流畅的空间设计。基地位置居于街廓转角，在周遭林立的建筑的包围下，设计师却能保有流动的光影变化，呈现出居家简约而宁静的氛围，如一座沉稳的建筑体量矗立于传统台湾街道巷弄中，同时享有独栋住宅的特色。

外观以一种现代沉静的手法，混凝土外墙间隔出车库区域，镂空形式导引进光线。夜晚，在间接照明的投射下，透出不同层次的光影。以大片落地窗恣意拥抱户外的暖阳，客厅和餐厅间以半透明玻璃砖区分空间领域，并引进部分室外光线，创造室内明亮氛围。

This is the home of a young couple and two children. Since the original pattern is out of use, when designer starts to set design planning to both structure dimenstion and interior, he takes more consideration of client preference, future lifestyle and site condition. The project is located in a corner of the street, embraced by surrounding buildings, designer succeeds in keeping floating light to creat a simple and quiet ambience, which made living in a building stand in traditional streets of Taiwan can also enjoy the free life just like living in a single family mansion.

The exterior design is in a morden way, spacing the garage area with cement wall. During the day time natural lights is spread into mansions from the meshed wall while during the nights, it take on a floating drawing by light and shadow from indirect lights. The interior space was divided by translucent glass bricks and large French windows which creat bright interior ambience.

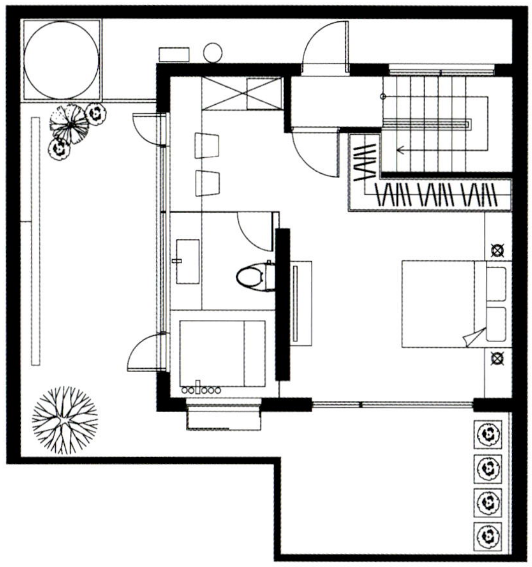

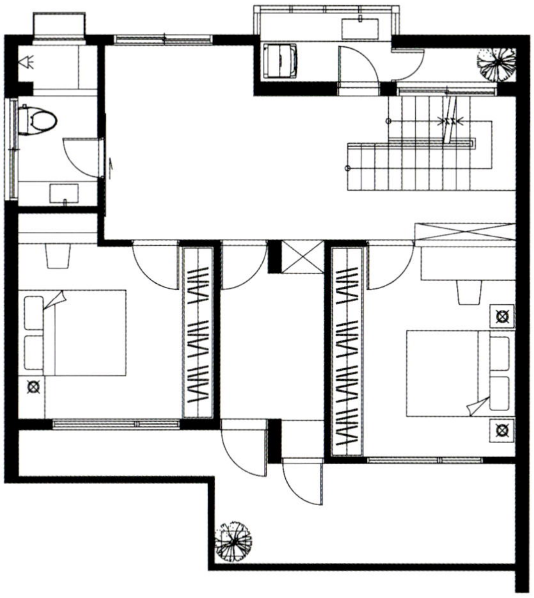

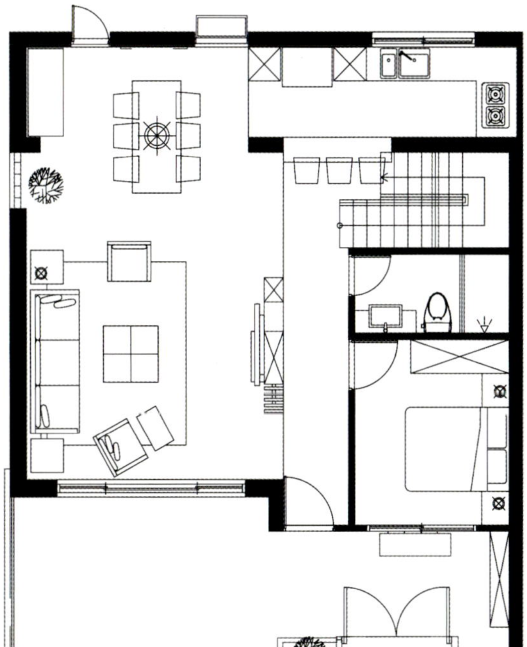

SHOREFRONT VILLAIN CBD

绿地望海 CBD 住宅样板房

项目名称 绿地望海 CBD 住宅样板房 / 设计师 颜呈勋 / 设计单位 Mrt Design
空间面积 300 m^2 / 项目地址 中国上海 / 摄影师 Moseman Eleanor Elizabeth

Project Name Shorefront Villain CBD / Designer Yan Chengxun / Design Company Mrt Design
Area about 300 m^2 / Location Shanghai, China / Photographer Moseman Eleanor Elizabeth

此样板房设计的重点是将原先两层的空间重新分割，改造成三层，并且保证了客厅的双层挑空，不失空间感。本案设计注重细节的刻画，恰到好处，为业主带来典雅、时尚的居住空间。在客厅里，咖啡色系的软包和墙纸、钛金板与灰玻结合的隔断及不锈钢材质的钓鱼落地灯，共同营造出一个整体、简约、温馨的空间。在二层的房间采用有通透感的玻璃作为隔墙，加强了空间的通透性。此外，设计师还利用走廊等小空间，增加了许多储物功能，给业主提供一个源于生活却高于生活的居住品质。

The design emphasis of the model house is redividing the space of original 2 storeys to 3, which keeps the double empty and also improves the sense of space. Great attention has been given to the details, bringing classic and fashionable living space to the future owner. In the living room, soft rolls and wallpapers of coffee color, the gray glass partition with titanium plates and a fishing floor lamp of stainless steel materials build complete, simple and comfortable space. The glass partition on the second floor enhances the transparency of the living room. Besides, the designer uses small space, such as the lobby, achieves great functionality, providing the owner a high-quality living space.

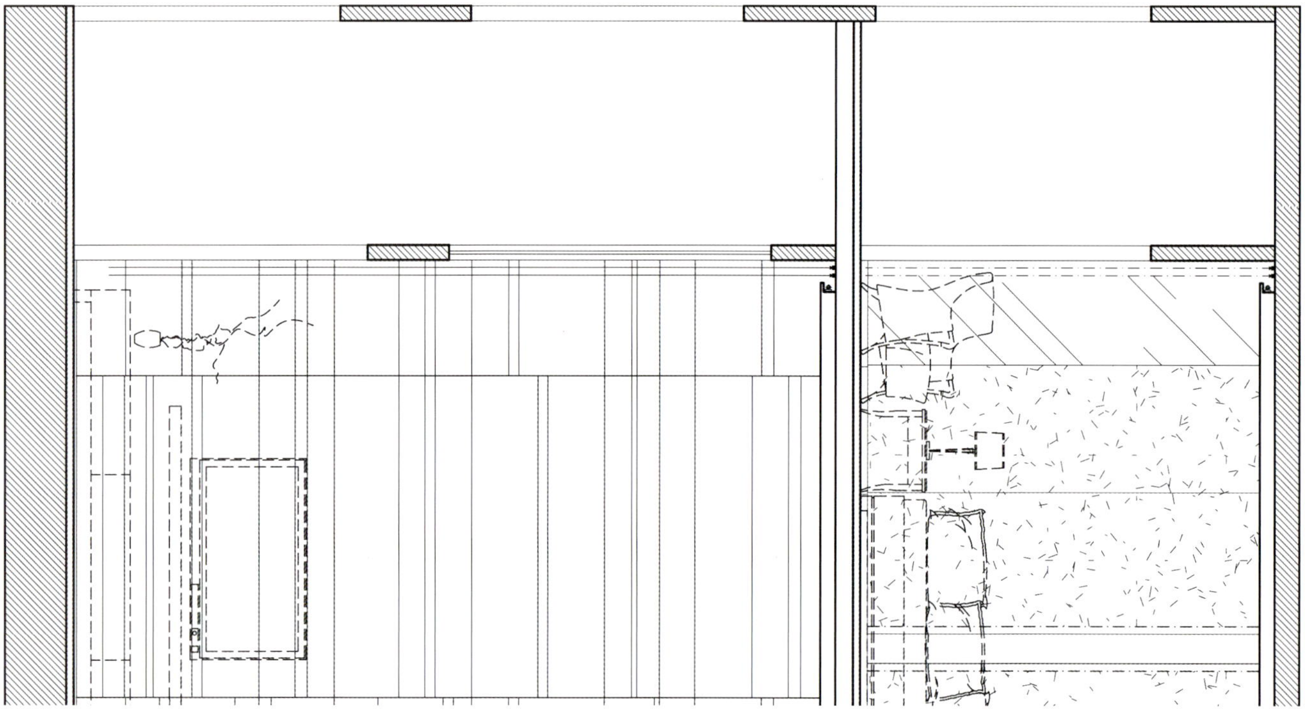

AGBARIA HOUSE
亚伯里亚住宅

项目名称 亚伯里亚住宅 / 设计单位 Ron Fleisher Architects / 空间面积 350 m²
项目地址 以色列 / 客户 阿德南·亚伯里亚 / 摄影师 夏·艾普什坦

Project Name Agbaria House / Design Company Ron Fleisher Architects / Area about 350 m²
Location Israel / Client Adnan Agbaria / Photographer Shai Epshtein

该房子位于穆斯木斯村里的河谷之上，在这个文化冲突频繁发生的地区，呈现出了多元的文化。业主要求建筑风格必须是当代的，同时不想摒弃陪伴他们成长的村庄记忆，特拉维夫的这家建筑公司对此重新诠释了巴勒斯坦式的建筑。房子的布局将传统的空间入口大厅与当代需求进行了结合，设置了电视机房及正式用餐厅。这种布局反映了屋主的愿望，在以色列社会中保持了独立的巴勒斯坦身份。

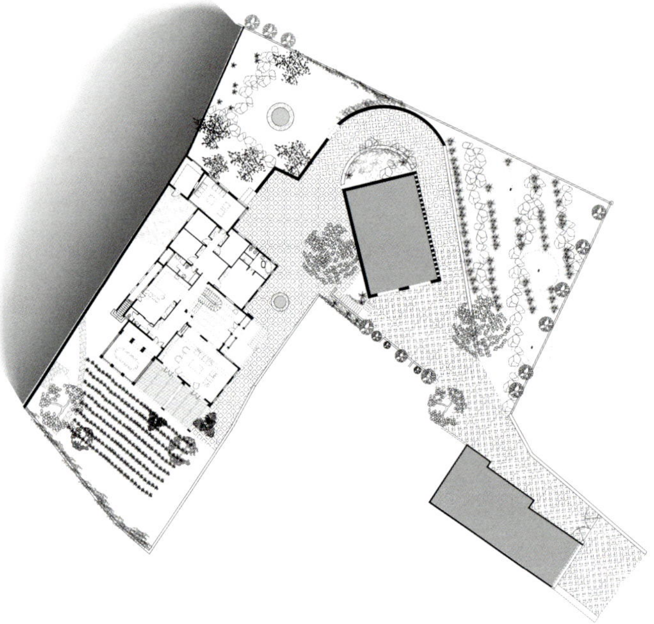

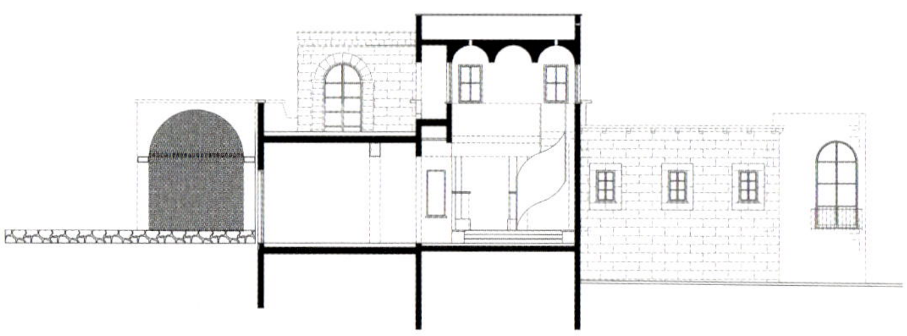

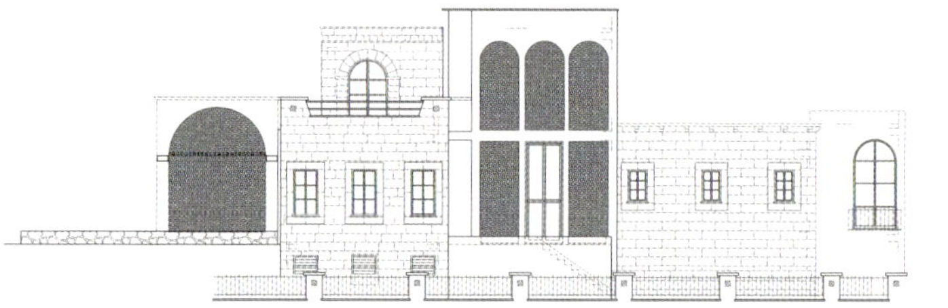

▶

房子位于山坡上，通往房子的大门则位于山脚下，距离房子有17 m的高差。在大门与房子之间，一条车道在农田景观之中盘旋而上。斜坡被传统的梯田分割，梯田由从橄榄树林中收集的当地石头筑成。车道环绕着白色的谷仓、石阶梯蜿蜒而上，通往顶部的房子，宽敞的露台可瞭望河谷风景。房子与古典穆斯林元素组成的自然景观及现代技术进行着对话。入口处的玻璃墙坐北朝南，以“瓦迪阿拉”图案作为底纹。入口大厅三道8 m高的拱顶门之上设置风扇进行通风，排出热空气，送来凉风。主客厅和餐厅直接通往屋外的花园，园中色彩鲜艳，被白色围墙包围。

In a region where cultures usually clash, the house over the "wadi"(valley) in the village Musmus is a multicultural experience. A cooperation between clients that asked for a contemporary architecture, but didn't want to forget the memory of the village they grew up in, and an architectural firm based in Tel Aviv created a reinterpretation of Palestinian architecture. The plan combines traditional spaces, as the "liwan"- the entrance hall, and contemporary needs, as a TV room, and a formal dinning area. It reflects the will to keep an independent Palestinian identity within the Israeli society.
The house is located on the top of a hillside overlooking "wadi ara". The main entrance to the property is more than 17m down the slope. Between the gate and main, house a driveway curves in a reconstructed agricultural landscape. The slope was divided with traditional terraces made from local stone collected in the families olive grove. The driveway surrounds the white barn, a staircase climbs to the top of the building to a wide balcony viewing the valley and welcoming the visitors into the private living area.
The house is in a dialogue with the natural landscape using classical Muslim elements as well as contemporary technology. The entrance glass wall facing south is shaded with an interpretation of a "Mashrabiya". The "liwan" is ventilated with passive suction through shutters located on top of 3 vaults 8m high. The hot air is sucked out and replaced by a cool breeze. The main drawing room and the formal dining room open to a walled garden ,colorfully framed by the white volumes.

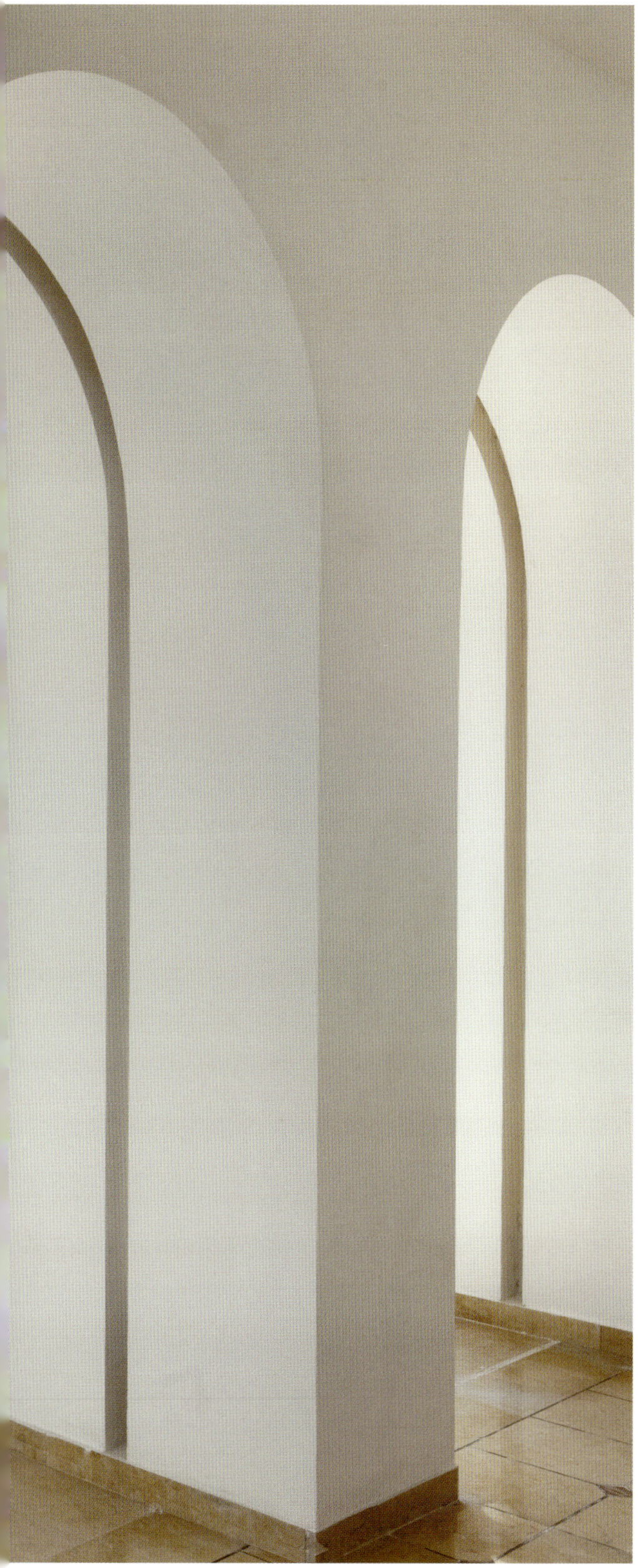

WU RESIDENCE, DANSHUI

▶ 淡水吴公馆

项目名称 淡水吴公馆 / 设计单位 春雨时尚空间设计 / 空间面积 281 m² / 主要建材 英式家具、复古地砖、陶瓷锦砖、镀钛、茶色玻璃、半抛光瓷砖、天然大理石、清玻璃、抛光石英砖 / 设计风格 新古典

Project Name Wu Residence, Danshui / Design Company Spring Shower Interior Design / Area about 281 m²
Materials Furniture In English Style, Retro Floor Tile, Mosaic Tile, Titanizing, Tawny Glass, Polished Tile, Natural Marble, Clear Glass, Polished Quartz Brick / Design Style New Classical

喜爱古典、优雅的英式家具的业主，在设计师规划室内空间之前，就与设计师先行为未来的居家空间挑选英式家具，虽然还没有任何设计草图出来，但带有强烈风格的英式古典家具就已清楚点出空间重点，设计师也依照家具塑造出来的空间氛围，为业主规划具现代精神的英式古典住家。

▶
The proprietor is a man who likes classical and graceful furniture in English style. Instead of firstly asking designer Jianzhi Zhou to plan the interior layout, he invited the designer to select furniture in English style with him for his future dwelling space. Although sketches haven't been done yet, classical furniture in English style with strong characteristics has clearly pointed out the emphasis of this space. And according to the ambiance the furniture brings, designer created this classical house in English style with modern spirit.

三组不同花色的英式沙发在客厅空间，相同的胡桃木边框及椅脚构筑古典、优雅的氛围，地板铺贴半抛光瓷砖，以其雾面的表层表达仿古精神，斜贴的砖面线条让空间感再度放大。暗红色绒布沙发后面是大型层架兼具展示与收纳功能，设计师在层架背景墙上贴饰红色图腾图案壁纸，丰富了空间色彩，营造出皇家贵族气质。为了突显英式家具的空间主体性，设计师将顶棚、柱体等线条简化。

简单线条营造层次感的手法，设计师也运用在客厅侧面的茶玻璃推拉门上，顺着入口处走入公共空间，设计师在右手边的线板茶玻璃后方，藏有大型收纳柜体，推拉门也可作为入口处的屏风，分隔内外不同空间，其穿透性的质感保证了入口处的通透明亮，收纳柜体的简单线条，也不会破坏设计师刻意突显英式家具的用心。当推拉门合上时，线条简单的喷砂线板成为室内美丽的端景墙。

3 types of sofas in English style with different designs and colors are situated in the living room, the same walnut frame and the feet of Gabrieli chair create classical and graceful atmosphere. Polished tiles are used for the floor, its mist-like surface presents archaized spirit, inclined stripes of tiles enlarge the sense of space. Behind the dark red charpie sofa, there's a big shelf which functions as both a display cabinet and a bookcase, designer adopted red wallpaper with totem for the wall behind the shelf, that increases space's chroma, creates a spectacular sense of royalty and nobleness. In order to feature the furniture, designer simplified lines of the ceiling and cylinders.

Simple lines to create a sense of layers, such representation is also used in the living room for a sliding door in tawny glass material, walking along the hallway into public space, you can see that designer placed a huge storage cabinet behind the tawny glass which is on the right side. When opening the tawny glass sliding door, the external and internal space is immediately separated, the see-through quality keeps the hallway bright and transparent. The simplicity of the storage cabinet still highlights the furniture in English style. When closing the sliding door, voile panel is changed into beautiful scenery wall for the interior space.

YANGMINGSHAN RESIDENCE
阳明山住院

项目名称 阳明山住院 / 设计师 刘嘉雯 / 设计单位 即[illegible]有限公司 / 空间面积 [illegible]
主要建材 进口复古地砖、进口壁纸、橡木染白、砂岩

Project Name Yangmingshan Residence / Designer Liu Jiawen / Design Company Jizuojizuo Interior Design Company, Area about 66 m² / Materials Imported Antique Tiles, Imported Wallpaper, Dyed-White Oak, Sandstone

坐落于台湾北部最美的阳明山，仅 75 m² 的约 34 年的旧屋要规划出三室二厅并含更衣室及两套卫浴的空间，强大的功能及穿透视觉的设计成为主轴。
客厅以投影布幕的设计取代传统电视墙面，巧妙地区隔了客厅与餐厅，使视野更加开阔，将台湾名山——阳明山的幽静山景收入眼帘，成为家的一部分，地坪以自然、浓郁的钟乳石地砖铺贴，沙发背景墙雅致的图腾壁纸，勾勒出空间表情，与户外碧绿景致相呼应。

▶
Located on Yangmingshan, the most beautiful mountain in the north of China Taiwan, the mere 75 m^2 old house was designed to have three bedroom and two halls with dressing room and two bathrooms. The high technology and transparent perspective become the main design concept in this case.

The living room is equipped with projector screen instead of traditional TV wall, sophisticatedly separating the space from the dining room and widening the space visually. The wonderful scenery of Taiwan's famous mountain is one of the distinct features of the residence. The floor is of natural limestone rock tiles, while the wall against the sofa is of very elegant totem wallpaper, depicting the expression of the space, matching the greenery view from the window.

▶
主卧内推拉门的设计，不但区隔了卫浴空间，也完整地营造了卧室温馨、休闲的氛围；主卧卫浴，除了运用奇零空间隔出干湿分离的区块，加上大面积窗户，完全将户外绿意纳入室内，让沐浴成为奢华的享受。
男孩房窗户采用格栅式设计，增加其采光效果，房内层架墙面与房外客厅层板展示柜设计相呼应，不失时尚感。女孩房拱形门的设计，在增加浪漫气息的同时也增添了森林小屋的童趣。
全室特殊纹理及透光的设计，通过不同格局的安排，无形中放大了视觉的空间，餐厅水晶吊灯及雅致的壁灯布置，更表现出家的温馨，结合周边环境借材取景，不枉山中美景，除兼顾时尚、休闲风格外，也具备强大的收纳功能，达到住的便利性，这些一直都是设计此房的中心理念。

▶
Sliding door is installed in the master room, not only dividing the bathroom and the bedroom but also creating a very warm resting space. Zigzag arrangement is applied in the bathroom to separate the dry area from the wet area, with 180° huge window frames, bringing outside greenery into the space and making bathing a luxury experience.
Window grids are used in the boy's room. The fashionable shelves design on the wall is conformed to the shelves design in the living room. As for the girl's bedroom, an arched door is molded to add a sense of romance and fantasy of a lovely forest log.
The special textile and light-transmittable design through the whole residence invisibly highlight the visual experience through different arrangement. The crystal chandelier and elegant lamps together create the temperature and hue of the house. Combined with surroundings' stunning landscape, the whole residence looks fashionable in a relaxing style and of high storage function, which achieves the residential convenience.

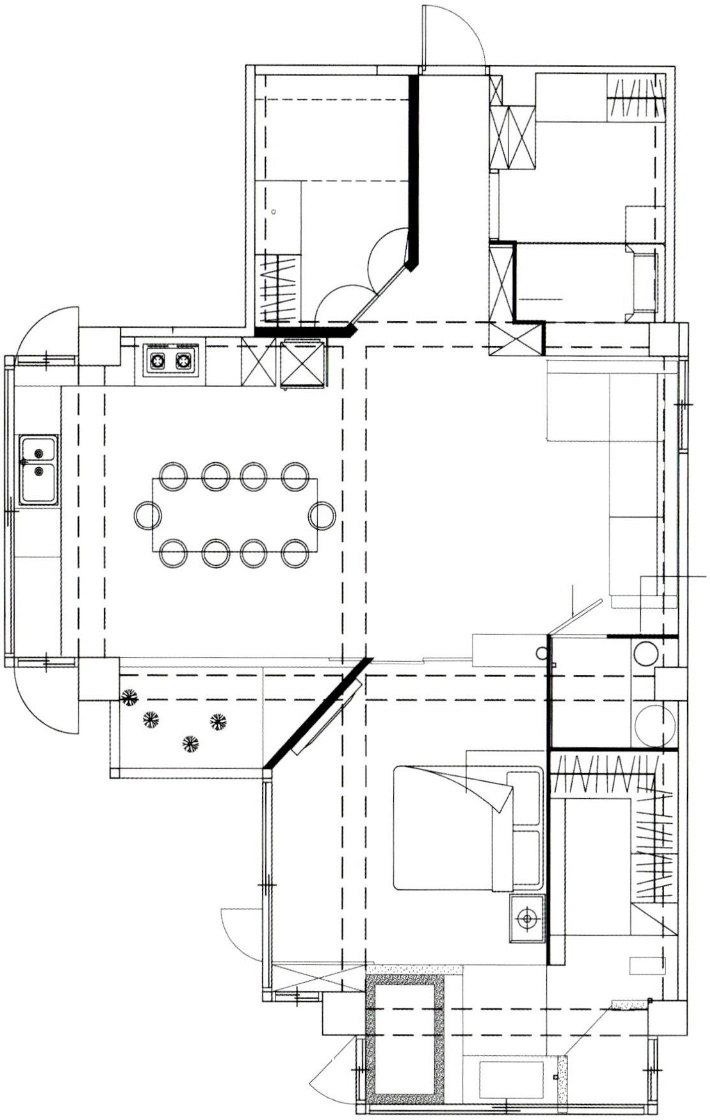

HOSANNA
山形和撒那

项目名称　山形和撒那 / 设计师　任萃 / 设计单位　十分之一设计事业有限公司
项目地址　中国台湾台北 / 作品类别　居住空间

Project Name　Hosanna / Designer　Ren Cui / Design Company　1/10 Design Company
Location　Taipei, Taiwan, China / Category　Residence

“和撒那”是希伯来语，在圣经的新约福音与四福音书中是赞美神的惊叹词。
本基地位于一林荫山边，业主为虔诚基督徒。男主人与女主人为了让每周的基督徒家庭聚会能有更宽广的空间，故将原本的四房改为两房，并把客餐厅、开放厨房、图书区作为最大区域来开放使用。
以山为形，以树为体。将四周环境的山形，引入室内，是为了隐藏贯穿客厅与餐厅中间的结构梁。随意停留在各角落的漂流木，是家中成员因着工作关系而在不同时期收集的摆饰，兼具家具与时间记忆的角色。每个停留在空间角落的漂流木，都有着恰如其分的命定。时而从天际线洒下的灯光，犹如山形与山形间相互辉映的曙光。光洁地面上倒影的间接光，犹如从山谷泄出的余晖，带给室内和煦温暖的照明。

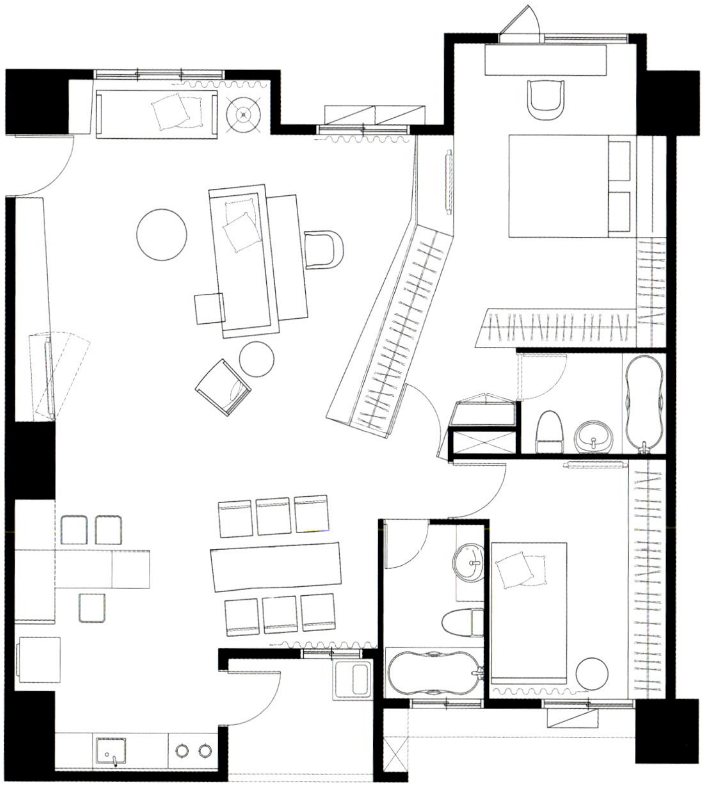

Hosanna is the word praising God in Bible.
The residence is located on the edge of a mountain and the owner is a devout Christian family. To have more space for family gathering held every week, the family couple decide to transform the existing 4 rooms to 2 rooms, with the guest restaurant, open kitchen and the reading area for opening use.
With the shape of a hill and a tree, the house enjoys the mountain view around, which is to hide the partition between the living room and the restaurant. The scattered floodwood in every corner is the collection of the family members, which plays the traditional role of decoration and also is a memory of past time. The smooth indirect light on the floor brings warm illumination to the interior, like the sunset glow which leaks out from the valley.

TIANMU DAI MANSION

天母戴宅

项目名称　天母戴宅 / 设计师　郭柏伸 / 设计单位　奇逸空间设计 / 空间面积　363 m²
主要建材　火山石、橡木、烤漆玻璃、人造石、铁件、玉檀香木地板

Project Name　Tianmu Dai Mansion / Designer　Guo Boshen / Design Company　Qiyi Interior Design
Area　about 363 m² / Materials　Volcanic Stone, Oak, Paint Glass, Artificial Stone, Iron, Jade Sandalwood Flooring

ANDO

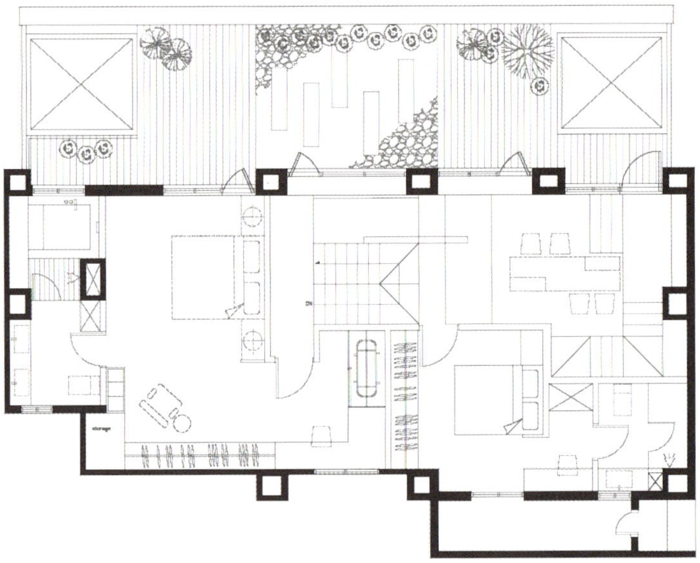

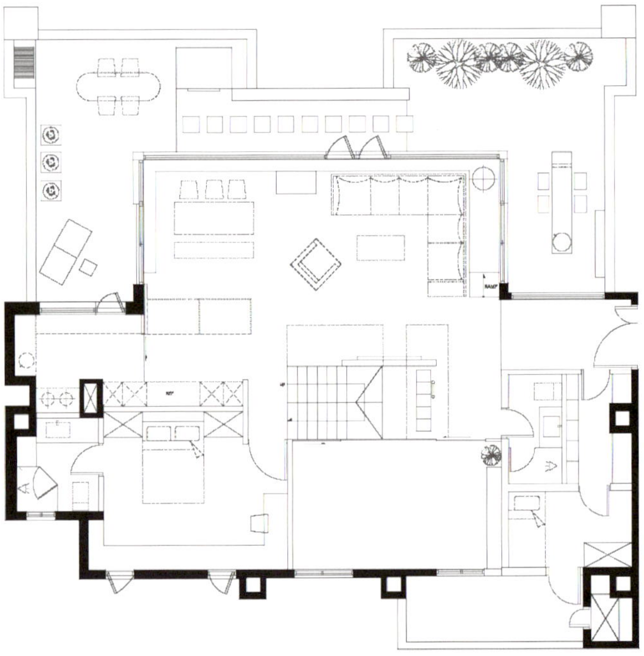

此案业主为国内知名服装设计师，其所成立的自创品牌在海峡两岸经营得十分成功，低调的她在居家生活方面十分讲究私密性，在空间舒适上也要求有一定的品质。

此两层楼住宅本身的优点是拥有大尺度的户外露台，可惜连接上下楼层的楼梯，却被挤在入口旁的拥挤角落，让上下楼层的生活空间关系总缺少了些什么。设计师大胆且清晰地提出自己的解决方案，包括迁移楼梯位置，并在保留梁柱的前提下，局部裁切楼板等。虽然工程繁复，却使两个楼层之间的关系顿时丰富而开阔起来。

首先，下方楼层主要为公共空间，拥有良好的视野，设计师充分运用此条件，将对外界面以开放式的露台与落地玻璃窗界定，然后餐厅、客厅以及户外吧台等沿着这个长轴展开，使之彼此连接。

整个公共空间以白色楼梯为背景，界定出功能清晰、动线流畅的生活空间。如入口左侧的鞋柜置物间、公用卫浴、客房与多功能活动间。连接垂直动线的白色楼梯以钢骨为结构，一阶阶地：往上蹿升，如一优雅的雕塑，一旁结合整面墙的书柜柜体，楼地板局部挑空后，上方走道以透明强化玻璃为地材，将上下楼层的视野串联在一起，让光从顶楼自由挥洒，打破了封闭的空间关系，舒展出上下穿透、层次井然的流动，使空间更加通透。

The client is a well-known fashion designer whose own brand is flourishing in both Taipei and mainland China. For dwelling, She tries to lay low, and cares for the private and free space of daily life. Besides, a comfortable and high-quality designed house is also required.

The advantage of this two-floor house is a large balcony, and the disadvantage is the narrow stairs neer the crowded entry. The audatious design solution contains stairs replacement and cutting part of floor. The large changes in work made the rooms more spacious and flexible.

The first floor functions as a public place. To make use of wonderful scenery outdoors, designers defined the front facade with opened balcony and French window. And the dining area, living room and outside bar connected with each other and stretched in the line of it.

The white stairs were taken as the background of public space ,and it also devided the living area functions such as entrance hall for cabinet, public bathroom,living room and multi-functional hall. Steel structure connects the suspended white stairs.Parts of the floor are cut out and instead of transparent toughened glass. Then the two floors are connected, light strolling in the whole house,which creates a penetrating and lightsome space as a whole.

HOLIDAY AT HOME

在家度假

项目名称　在家度假 / 设计师　总监 丁荷芬、冯慧心 / 设计单位　采荷室内设计 / 空间面积　264 m²
主要建材　西班牙复古陶砖、手染松木、实木板材、南方松实木、锻铁、抿石、化石、红砖、琉璃珠、瓷砖、水晶灯具、家饰布等 / 项目地址　中国台湾台北

Project Name　Holiday at Home / Designers　Ding Hefen, Feng Huixin / Design Company　Caihe Interior Design / Area　about 264 m²
Materials　Spanish Retro Ceramic Tiles, Hand-Dyed Pine Wood Boards, Southern Pine Wood, Wrought Iron, Smooth Stone, Natural Fossils, Red Brick, Glass Beads, Tile Collage, Teak Wood Furniture, Crystal Chandelier, Furniture Fabrics, Etc / Location　Taipei, Taiwan, China

此设计案为中国台湾台北的别墅，设计总监丁荷芬以开放式的手法，引进良好光线，结合中国风水与西洋色彩能量学，打造这间时尚的居住空间。客厅公共区域采用开放式整体设计，带来无压力的空间感，顶棚的实木条和墙壁石材的使用，营造出乡村风格，同时有视觉导引作用，仿古砖也衬托出空间的舒适质感。

厨房中的手工实木橱柜和以手染木质颜色的台面及西班牙进口陶砖，共同营造出温馨的氛围，实现了女主人对于厨房的梦想。私人空间卧房的设计，总监丁荷芬善用寝具及家饰布搭配，将空间营造得舒适又优雅，置身其中，能够使人放松一整天的紧张与疲劳。

▶
This case is to design a villa in Taipei of China Taiwan. The design director Ding Hefen adopts the open approach, combining Chinese Fengshui knowledge and the study of western color energy, to create this modern villa of country style. The absence of wall in the public area allows the light coming from each window to shine round the space. The hardwood beams on the ceiling and the stoned wall together bring a vibrant visual experience of country style, with retro floor tiles adding cozy sensation to the space.

Hardwood kitchenware, hand-painted Nakashima workbench and Spanish imported terracotta are used in the dining room and kitchen to realise the mistress' dream of a kitchen. As for the bedroom, the Design Director Ding Hefen creates a relaxing and elegant environment through smart arrangement of the decorations and selection of bedding.

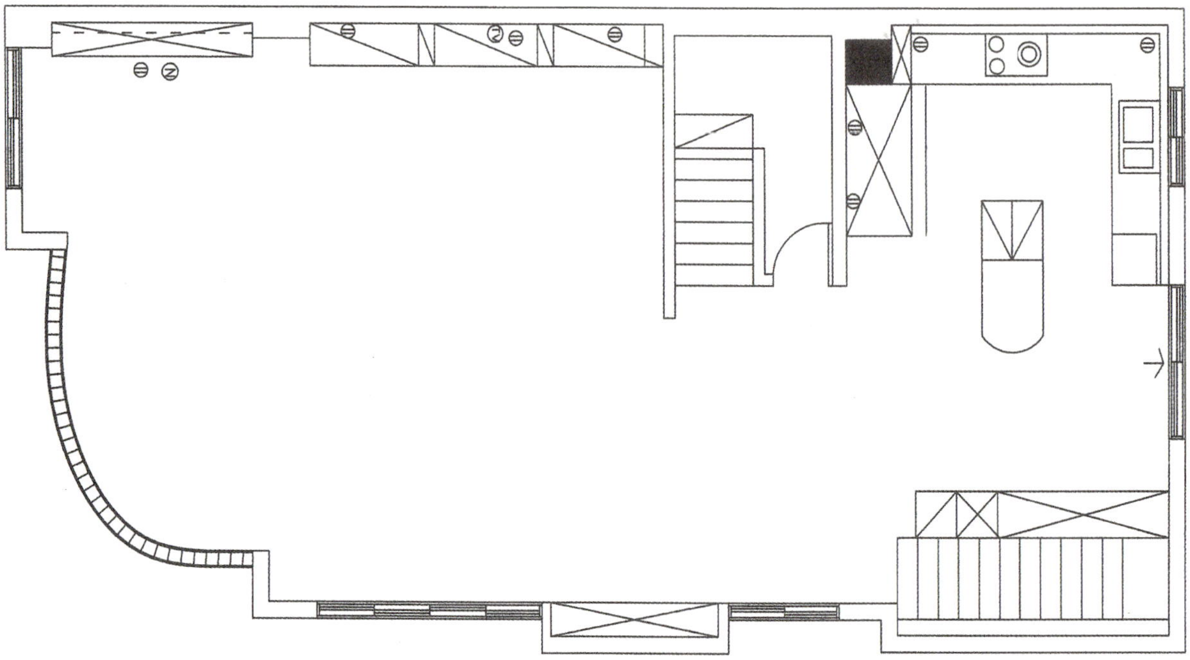

囍

ZHONGSHAN VANKE LANGRUN VILLA GARDEN

万科中山朗润园别墅样板房

项目名称　万科中山朗润园别墅样板房 / 设计师　韩松 / 设计单位　深圳市昊泽空间设计有限公司 / 空间面积　277 m² / 项目地址　中国广东中山

Project Name　Zhongshan Vanke Langrun Villa Garden / Designer　Han Song / Design Company　Shenzhen Gaoze Space Design Limited Company / Area　about 277 m² / Location　Zhongshan, Guangdong, China

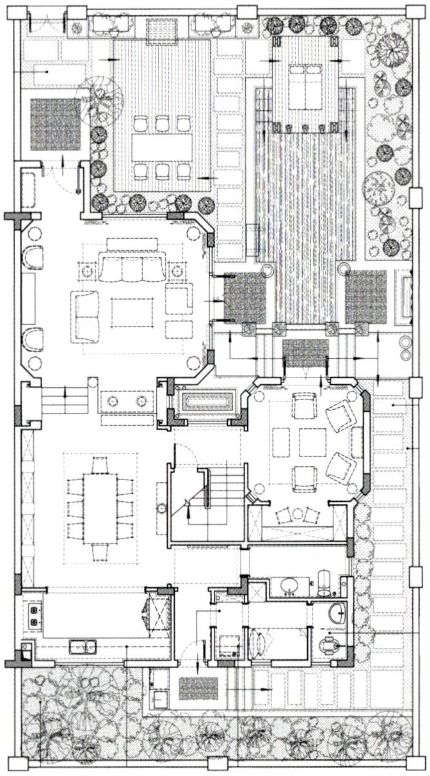

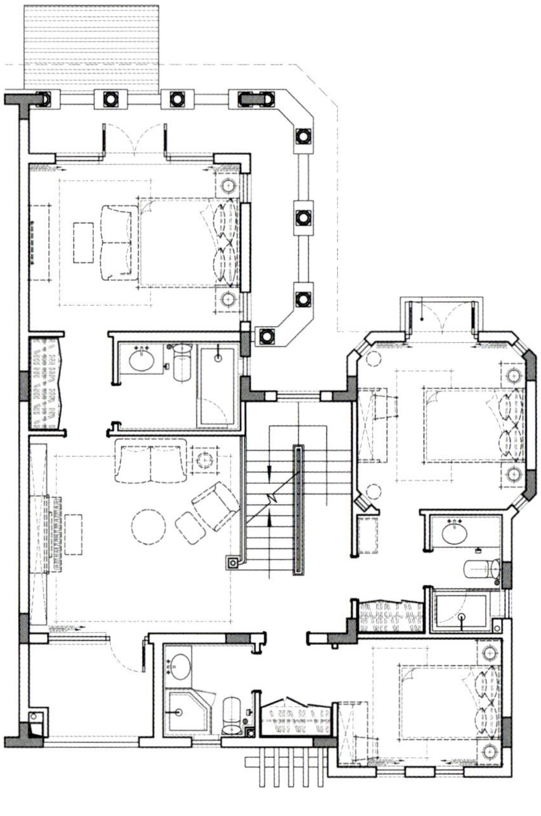

▶

每个人心中都有个理想国，业主真正需要什么？

一、首先是更贴心、更真实全面、层次丰富的功能节点 。向业主传递一个强烈的信息：我的设计是真正适合居住的，是真正满足业主居住生活理想需求的；

二、强化空间感，增加通透感和可拓展性；

三、追求古典的均衡对称美，以及由轴线所带来的丰富的空间层次感；

四、与户外园林景观的互动与对话；

五、我们给的是一种生活方式——精神层面的东西，它应该具备以下气质：

a. 功能 b. 安全 c. 静 d. 思辨 e. 自然 f. 健康

我们做的其实是把业主内心的真实需求释放出来，并且扩大和发散。

Inside everyone's heart, there is a Utopia. What do our clients really need?
First, function nodes. They need functions that are more caring, complete and varied, and convey to the clients a powerful message: our products are the real residence-friendly houses, and the real satisfaction to the consumers who live in them.
Second, fantastic sense of space, transparency and flexibility in the space.
Third, classic aesthetic sense of balance and multiple spatial layering.
Fourth, interaction and dialogue between the house and the gardenscape.
Fifth, a spiritual lifestyle should contain these qualities: functional, safe, quiet, dialectic, natural and healthy ones.
What we do is actually to realize and enlarge the clients' ideals in their minds.

WANTAI SPRING COMPOUND MANSION

万泰春天复式住宅

项目名称 万泰春天复式住宅 / 设计师 李增辉

Project Name Wantai Spring Compound Mansion / Designer Li Zenghui

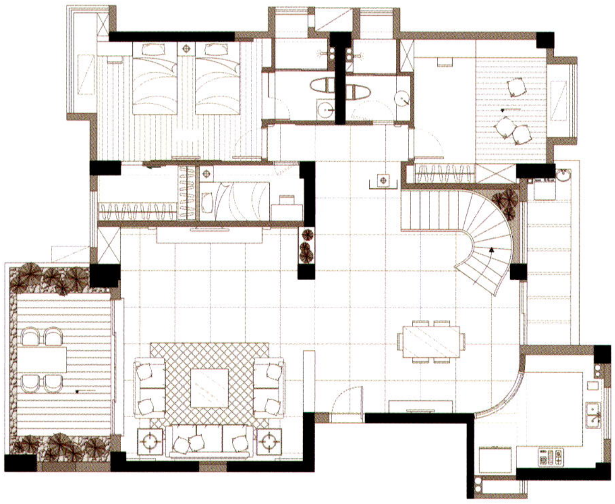

▶

由于本案极好的户型条件，且每个功能空间都有较好的独立性，采光良好，外景迷人，因此，空间轴线和外景摄取成为室内设计的主线。
在设计风格上，主要以现代、时尚为主，线条简洁，格调明快，材质的选用丰富多彩，力求打造不同个性的艺术空间，色调的选择则强调自然、明朗，空间布局合理、精练。

The layout of this case, the functional independence of each space, good lighting and superb scenery have made it possible to create a central axis in the house and take advantage of the window view.
The design is of modern fashionable style. The application of simple lines, fresh tone and choice of multi-colored materials aims to produce a unique artistic space with natural quality and sophisticated layout.

FUNK

MODERN ARISTOCRAT
摩登新贵

项目名称　摩登新贵 / 设计师　李光政 / 装修公司　DOLONG 董龙设计 / 空间面积　260 m^2
软装材料　手工绣花窗帘、进口壁纸、水晶吊灯 / 硬装材料　罗丹皮纹砖、蜜蜂仿古砖、仿古地板、金属马赛克、杜拉维特、洁具、威乃达橱柜

Project Name　Modern Aristocrat / Designer　Lee Guangzheng / Company　DOLONG Design / Area　about 260 m^2
Soft Outfit Materials　Handmade Rust Spend Curtain, Imported Wallpaper, Crystal Droplight / Hard Outfit Materials　Rodin

源于对最为简单生活的追求，富足的生活之外，人们的内心还有自己的梦想，本作品就这样应运而生。设计师延伸了建筑设计精神，充分利用户型的特点，将保姆房拆除并融入厨房，在拓展空间视觉的同时，解决了厨房的功能性问题。让室内空间延伸并与室外环境相融合，营造出舒适、通透、大气的空间。客厅、餐厅以及厨房流畅、通透的空间关系，表达了富有活力的生活状态。利用合理的功能划分，对称和简练的设计元素，来体现家居的美观性。通过咖啡色仿古砖、白色实木雕花、皮纹砖，茶镜以及金属马赛克等华丽的材质和精致的家具搭配，从而散发出华贵与时尚的气息，这是新贵阶段在文化与艺术追求上结合的高品质住宅。

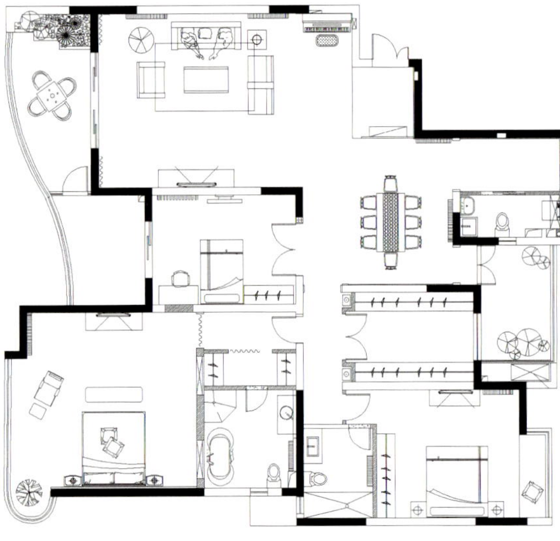

▶

Besides the simpe dream of rich life, people still have other pursuits, which are the theme of the project. The designer expends the architecture design spirit, taking full advantage of the space, removes the nanny room and blends it in the kitchen, which expands the space vision and enhances the practical applicability. The interior space is intergrated with the exterior environment, presenting a comfortable, transparent and generous space. The smooth and transparent space relations between living space, dining room and kitchen manifest the dynamic of life. With the reasonable space plans, design elements of symmetry and simplicity bring the beauty of modern houses. Coffee classic bricks, white real wood carve patterns, skin stones, tawny glass and metal mosaic, the luxuriant materials and exquisite furniture throw off luxury and chic. The new aristocrats love high-quality residences intergrating culture and art.

SOUTHEAST ASIAN CHARM

东南亚风情

项目名称 东南亚风情 / 设计师 张谨 / 设计单位 武汉梵石艺术设计有限公司
空间面积 170 m² / 项目地址 中国湖北大冶

Project Name Southeast Asian Charm / Designer Zhang Jin / Design Company Wuhan Fanshi
Area about 170 m² / Location Daye, Hubei, China

本案以现代手法诠释东南亚的装饰元素，把东南亚风格中的浓墨重彩、原汁原味地凸现出来。业主选择代表着奢华、唯美的泰式家具，所以全屋以泰柚木色为中心，显得神秘而又华贵。

The project is a good annotation of decoration elements of Southeast Asian-style with modern skills, in which the thick and heavy colours featured in Southeast Asian-style are presented in an uncut way. The client needs the luxuriant Thai-style furniture so that the Thai teak color dominates the house, which makes it mysterious and elegant.

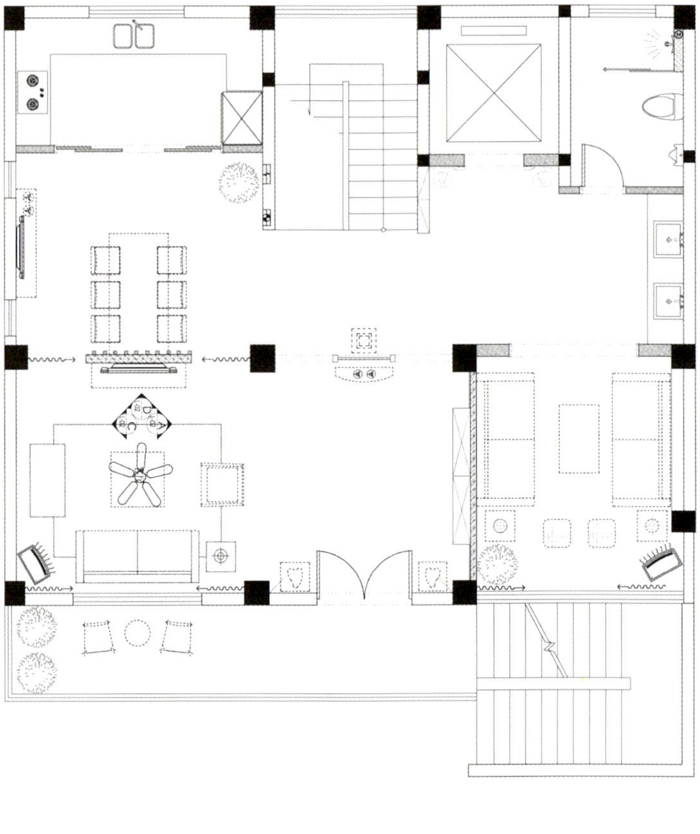

东南亚风格的空间一般色彩比较浓烈，如果不注重元素的合理搭配，就会使得空间繁复累赘。本案为了避免出现这一缺陷，充分利用灯饰和家具加以配合，并且巧妙地运用隔断作为装饰，大气而不失活力。

The color of Southeast Asian style is usually strong. Without a proper collocation of elements, the space will become heavy and burdensome. To avoid this, the designer uses decorative lighting with furniture, with a proper application of partitions as decoration, making it elegant and dynamic.

HACIENDA SUMMER HOUSE DESIGN CONCEPT

Hacienda 避暑别墅设计概念

项目名称 Hacienda 避暑别墅设计概念 / 设计单位 Mohamed Fares, Alchemy 设计工作室
空间面积 190 m² / 项目地址 埃及亚历山大北部海岸 / 摄影师 Sherif Tamim

Project Name Hacienda Summer House Design Concept / Design Company Mohamed Fares, Alchemy Design Studio
Area about 190 m² / Location North Coast, Alexandria, Egypt / Photographer Sherif Tamim

▶
奢侈而引人注目，这个 190 m^2 的豪华避暑别墅的各个空间都被设计为一个私人的避难所。把主要空间分为不同领域，不同的功能，使它成为招待客人的理想服务空间。
整个房间将不同的材料和颜色混合在一起，通过粗糙和光滑的纹理、冷暖色调以及各种面料混杂创建了一个协调体。超大沙发与黄金器皿搭配，为木制的摩卡背景添加一丝现代豪华的视觉冲击，此处为设计师最喜欢的空间。天然棉的超大沙发被放置于天窗下，营造出放松身心的理想场所。

Lavished with attention; each corner of this 190 sqm luxury summer house has been designed to create a private sanctuary. The main space is divided into different areas having different functions also making it ideal for entertaining guests.

The room as a whole mixes different materials and colors together, in a synergy created by the mélange of rough and sleek textures, bright and cool colors as well as a variety of fabrics. Described as the designer's favorite space in the house, the gold pots with the oversized sofa add a twist of modern luxury against the wooden mocha backdrop. The oversized sofa, with its natural cotton textile, is perfectly placed under a skylight to make it a perfect spot to relax and unwind.

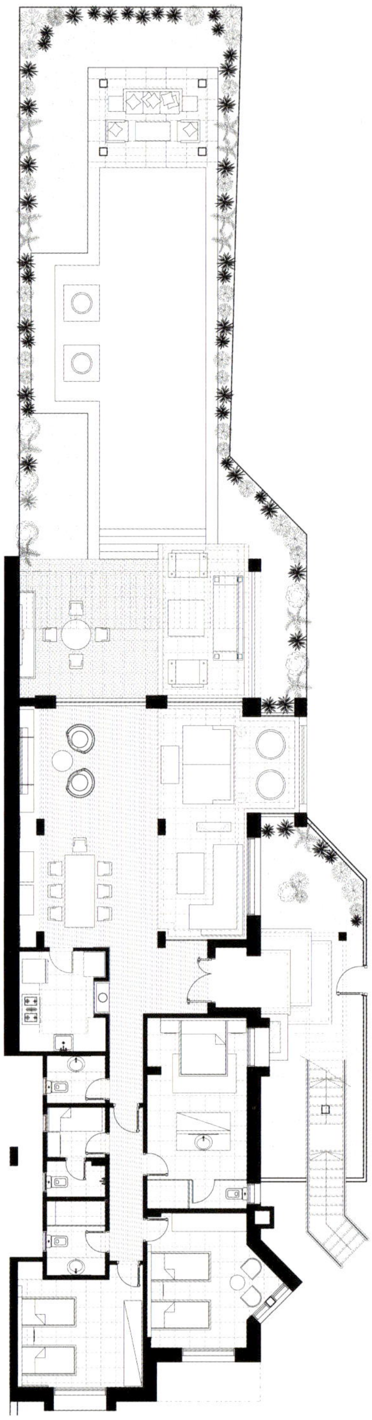

室外区域和游泳池被设计为别墅的自然延伸，由一个平台贯通全屋。环绕游泳池的露台与户外区域被设计成夏日芬芳的私人避暑胜地。这个鲁滨逊·克鲁索的空间让人感觉就像是走入了另一个世界，洒脱的背景、粗犷的纹理搭配着石板、大理石舒缓的瀑布声，为空间增添了一份独特的民族气息，更加使人仿佛置身私人避难所之中。

With a platform that runs throughout the house, the outdoor area and swimming pool were designed to be a natural extension of the villa. The terrace, which envelops the pool, is a private retreat to spend balmy summer days. This Robinson Crusoe space is like stepping into another world, an unstructured setting with rugged textures in addition to the soothing sound of the slate/marble waterfall, adding a distinctive touch of ethnicity to this private refuge.

Show Flat at Zhongxin Mangrove Bay 28-9 Villa, Zhuhai

珠海中信红树湾 28–9 别墅示范单位

项目名称　珠海中信红树湾 28–9 别墅示范单位 / 设计师　邱春瑞 / 空间面积　320 m²
主要建材　意大利木纹、山西黑、爵士白、黑金花、银白龙、雨林棕、青古板、金蜘蛛、罗马洞石、黑白根、火山岩饰面板、实木地板、陶瓷饰砖 / 项目地址　中国珠海香洲前山大桥侧

Project Name　Show Flat at Zhongxin Mangrove Bay 28-9 Villa, Zhuhai / Designer　Chiu Raynon / Area　about 320 m²
Materials　Italian Wood-Pattern Floor Marble, Shanxi Black Stone, White Jazz, Portoro, Silver Dragon, Rainforest Brown Marble Stone, Green Antique Stone, Golden Spider Marble Stone, Travertino Romano, Black Marquina, Effusive Rock, Wood Veneer, Hardwood Floor, Mosaic / Location　Beside Qiangshan Bridge, Xiangzhou District, Zhuhai, China

▶
本案位于珠海市香洲区前河西路。现代人的生活情结在空间中展示了出来，不同的空间带来的是不同的生活品质和生活氛围，人们希望由空间这种外在的物质形态给生活带来看不见的、内在的精神。空间的精神如同月亮之于夜空，因有精神，才凸显出空间的美好情怀。
设计师借助多种不同的元素来打造空间的每个视觉点和面，在不同的视觉结构中，力求凸显出空间的完整性和饱满感。客厅中将光线进行很好的“收藏”，凸显出空间的明暗对比感，同时也增强空间的张力感。意大利木纹带来的不仅仅是清晰和时尚，一同散发、萦绕在空间中的，还有几分空间于业主头脑中所形成的的清新感想和对生活的美好向往。客厅中主要以皮质沙发、玻璃桌、灰白沙发抱枕的色彩为基调，在此形成了适合业主特性的稳重感和低调的奢华感。

Located on the Hexi Road of Xiangzhou District in Zhuhai, this project presents modern people's preference of lifestyle. Each space brings different living quality and ambience. By this outer material forms, human beings experience their invisible inner transience. The spirit within the space is like the moon in the night. It is the spirit that makes the spaces nice. Designers applied a variety of elements to create each visual point and each side, which ultimately form the completeness and richness of the spaces. The lights in the living room are hidden sophisticatedly to promote the sense of space with shades of light. The use of Italian wood-pattern marble on the floor does not only lend the clarity and modernity to the space but also reflects the fresh thoughts and positive prospects of the house owner. The living room is toned with leather sofa, glass table and light gray cushions, revealing the owner's composed, humble and dignified characteristics.

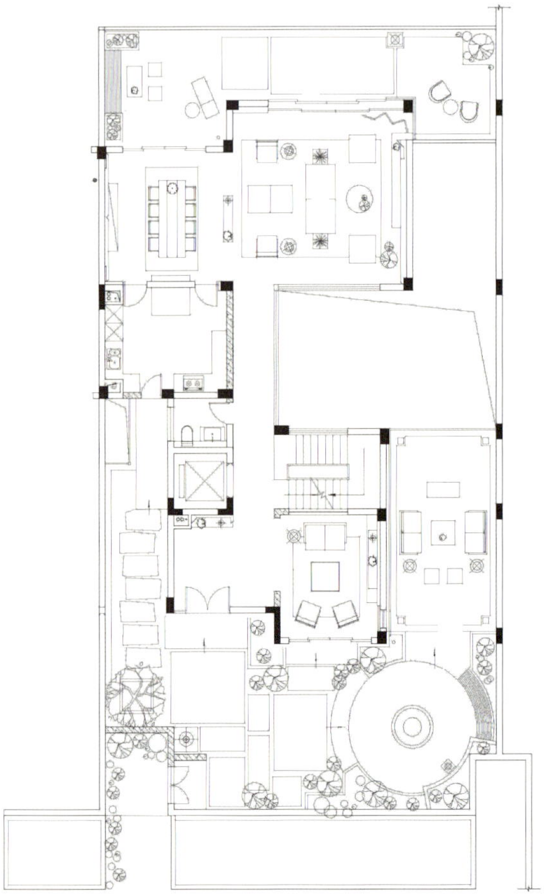

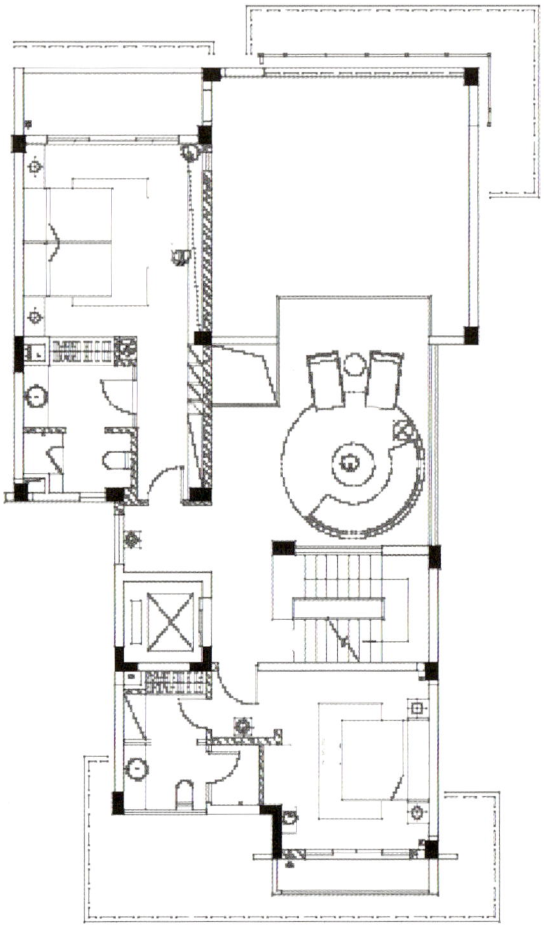

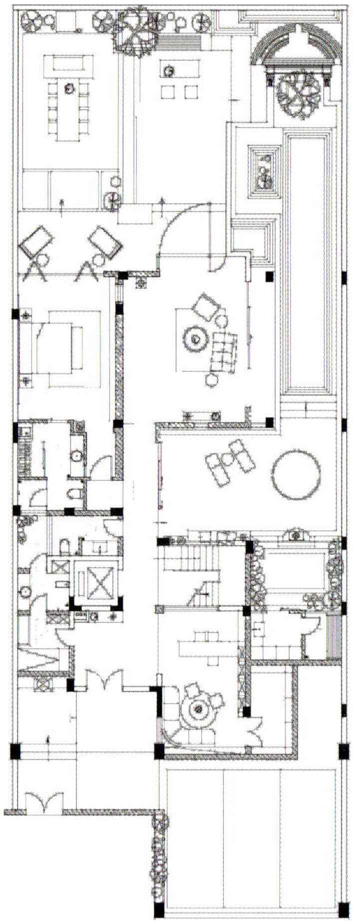

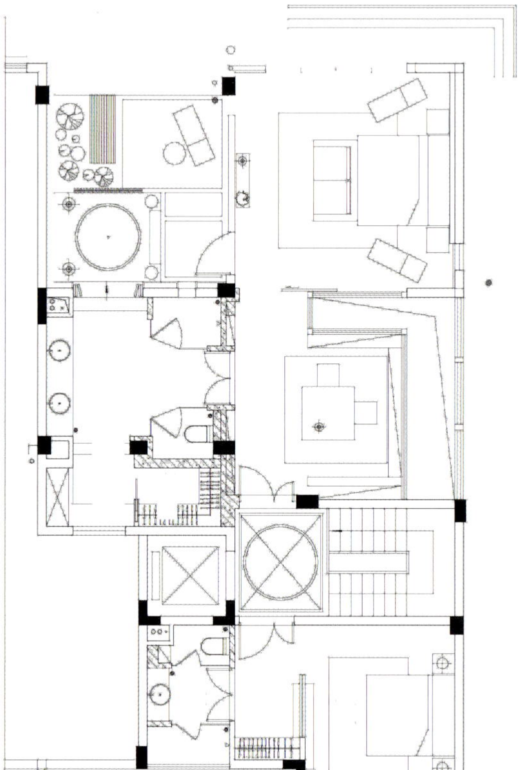

VILLA OF HESHENG INTERNATIONAL CITY

合生国际城别墅某宅

项目名称 合生国际城别墅某宅 / 设计师 周少元、周少朋 / 设计单位 宁波宫湖建筑设计工作室
空间面积 300 m² / 主要建材 仿古美式艺术砖 、美式家具、沙比利饰面板做旧、进口壁纸、复古浮雕地板、云石灯

Project Name Villa of Hesheng International City / Designers Shaoyuan Zhou, Shaopeng Zhou / Design Company LPI / Area about 300 m² / Materials Archaized American Art Tile, Furniture In American Style, Sapele Veneer, Imported Wallpaper, Retro Embossed Floor, Marble Lamp

本案例属于别墅豪宅设计类别，面积为 300 m²。别墅地处日益发展的镇海新城南部核心区，自然环境优美。此案内部进行了重新规划设计，作为亚别墅，原布局结构较差，经过重新设计改造后，客厅视野得到了扩大，主卧朝南，与阳台融合，空间上分设地下半层、地上三层，前有院子、后有花园，通风采光好，户型方正，结构合理，内部改造设计使整个别墅焕然一新，既落落大方又不失文化底蕴。

This case can be classified as luxurious villa design, the project area is 300m². The villa is located in the center of the south of Zhenhai new town, which is increasingly developing, and the environment there is beautiful. The interior has been planned and designed all over again, previously as a second-level villa, the original structure was comparatively a lot worse. After reconstruction, the visual field of living room becomes wider, master bedroom faces the south side and harmonizes with the balcony. As for spatial layout, a half storey below and 3 storeys above. Besides, an unoccupied area in the front and a garden at the back. Both ventilation and day lighting are good, unit of the house is nice, structure is proper. Reconstruction and redesign of the interior make the villa brand new and fresh, natural and graceful, and still keep the cultural core.

▶

本案在设计手法上，突出了文化人温文尔雅、平和理性的特点，用浅咖啡色的整体色调，表达出业主对温馨、典雅生活的向往。在设计风格定位上，吸取了文艺复兴时期巴洛克风格中的一些经典元素，既不过分张扬，又恰到好处地把雍容华贵之气渗透到每个角落，既突出了别墅本身的自然优势，又适当地彰显出业主的个人品位。

The design method shows that educated people are gentle and polite, also amiable and rational. Light brown is the main color tone and expresses elegance of the proprietor. When it comes to design style, designer absorbed classical elements of Baroque which belongs to the renaissance period. It's not like overly showing off. It appropriately sprays grace and nobleness everywhere. Furthermore, it highlights villa's self-advantages and proprietor's personal taste.

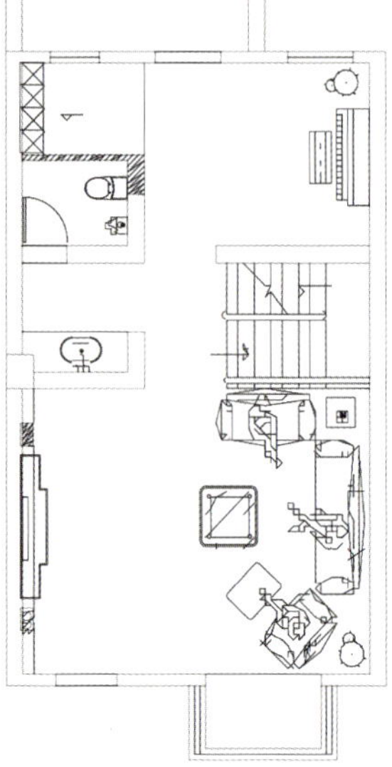

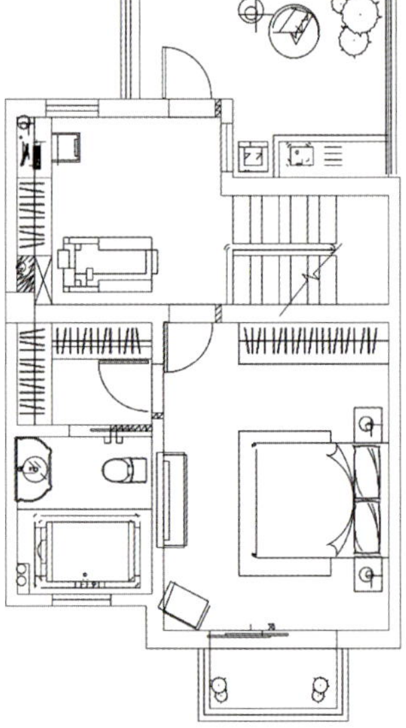

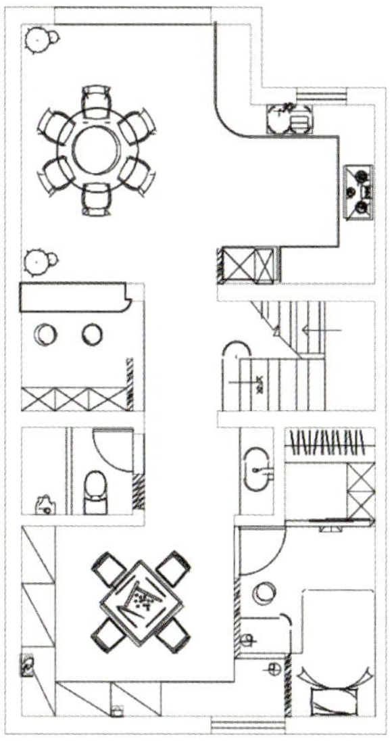

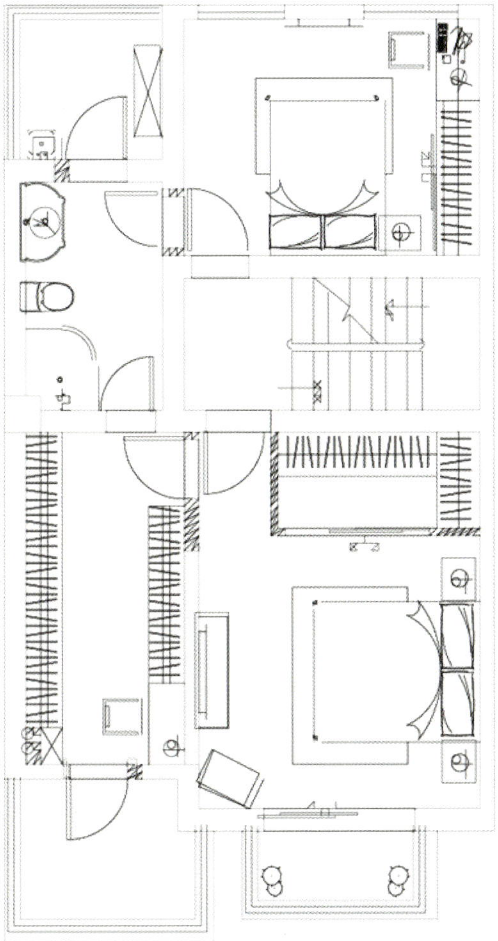

Residence of Spring River Floral Town

春江花城某宅

项目名称　春江花城某宅 / 设计师　周少元、周少朋 / 设[illegible]宫湖装饰 .LPI 别墅臻品设计工作室
空间面积　280 m² / 主要建材　亚振家具、水曲柳饰面[illegible]口壁纸、复古浮雕地板、ICI、砂岩 / 项目地址　中国浙江宁波

Project Name　Residence of Spring River Floral Town / Designers　Zhou Shaoyuan , Zhou Shaopeng / Design Company　LPI
Area　about 280 m² / Materials　A-Zenith Furniture, Fraxinus Mandshurica Panel, Imported Archaized Brick, Imported Wallpaper, Retro Embossed Floor, ICI, Sandstone / Location　Ningbo, Zhejiang, China

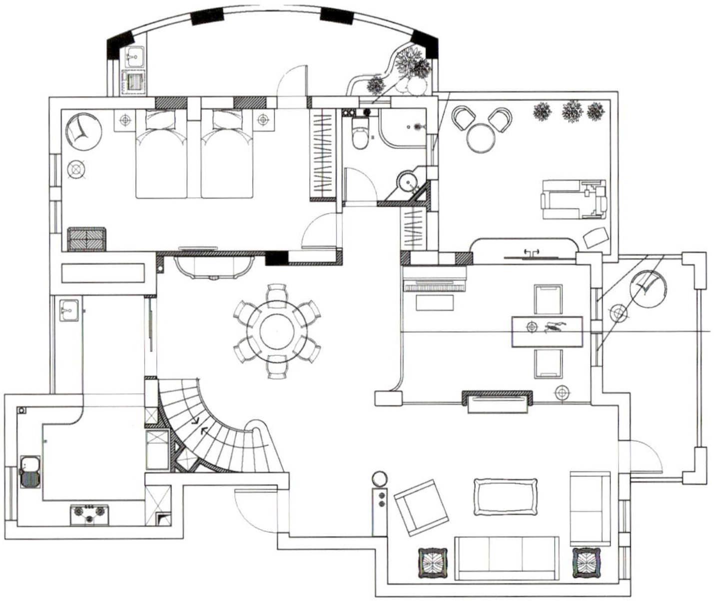

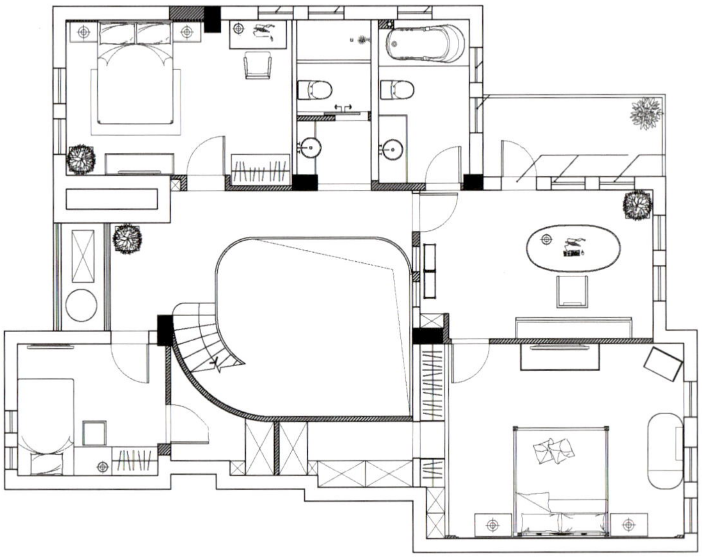

▶

这是三世同堂的居住空间，与业主沟通后，空间定位为欧式古典风格，并以乡村韵味点缀。 结构上，餐厅以弧形沿墙而上，打破了原有格局，使餐厅空间更显大气。布局上，一层考虑对外功能以及老人家的方便、舒适与休闲；二层的设计主要体现业主的舒适生活，儿童房的地中海风格、随处可见的绿植、尽享天伦之乐的照片墙、充满整个居室空间的装饰品以及油画！

▶

This dwelling space is for one big family formed by 3 generations. After communicating with the proprietor, we decided to make it classical with European style ornamented with village aroma. The dining room is in an arc-like structure which climbs along the wall, this is really a breakthrough. It makes the space more spectacular. As for the first floor, we planned the external function and considered convenience, comfort and leisure for the elderly; when it comes to the second floor, it mainly expresses cozy life of the proprietor, Mediterranean style of child room, landscape that can be seen everywhere, a photo wall showing endless happiness, adornments, oil painting. All these bring warmness to the living room.

No.2 Show Flat of European Style at Jingui Mansion

金桂华府样板房二号楼

项目名称　金桂华府样板房二号楼 / 设计师　邢新华 / 空间面积　131.81 m²
主要建材　人造石、灰镜、瓷砖、壁纸、木线条 / 项目地址　中国广东东莞

Project Name　No.2 Show Flat of European Style at Jingui Mansion / Designer　Xing Xinhua / Area　about 131.81 m²
Materials　Artificial Stones, Gray Glass, Ceramic Tile, Wallpapers, Quarter Round / Location　Dongguan, Guangdong, China

本案针对有一定经济基础的人群进行设计，本着典雅、时尚的理念，富贵而不乏亲和、典雅而不显做作、温馨而不失个性。所有的造型、灯光、配饰都将在这个空间的“家”的感觉中无限地放大，让人觉得处处皆有亮点，但又全部融合在一个整体中。
将欧式元素在空间中很好地运用，以现代的设计手法将欧式的沉重和烦琐去掉，把新古典主义运用到极致，将人带入到一座现代的“宫殿”。

卫生间
厨房
餐厅
卧室
客厅
阳台

卫生间
卧室
儿童房
客厅中空

▶

The design is for the people with fine economic background. With the concept of elegance and fashion, the house is designed to have glory and affiliation, being classic and natural, snug and creative. All the decorations, lights and accessories used stress the idea of "family" and leave lightspots everywhere and also they are intergrated into the whole house. European style elements are in good use. With the modern design skills, the designer removes the heaviness of the original European style, uses the neoclassicism, building a modern palace.

Zhongxin Mangrove Bay Villa, Shenzhen

深圳中信红树湾别墅

项目名称　深圳中信红树湾别墅 / 设计师　孙昌文 / 空间面积　380 m²
主要材料　西班牙米黄、榉木喷白漆、意大利瓷砖 / 项目地址　中国广东深圳

Project Name Zhongxin Mangrove Bay Villa, Shenzhen / Designer Sun Changwen / Area about 380 m²
Materials Cream Marfil, Beech White Paint, Italian Ceramic Tile / Location Shenzhen, Guangdong, China

本项目位于深圳红树湾区，是中信集团又一个成功的项目。设计采用环保材料以及进口瓷砖、意大壁纸、榉木喷白漆、实木地板等。原有平面结构比较零散，建筑外观比较现代，色彩为单一的冷灰色。设计师多次与甲方沟通，在风格定位上有很多争议的地方。设计师认为室内设计是为居住者服务。房子设计再另类、个性、前卫，那也只是展示。必须以人为中心。所以在设计上打破单一美式设计风格。结合中国特有的红色，在美式大气、稳重中，又不失温馨。

Located in Hongshuwan, Shenzhen, the residence is another success of Chinatrust Group. Eco-friendly materials are applied, such as imported tiles, Italian wallpapers, beech white paint and hard wood floor boards. The original plan structure is some kinda scattered, with the facade modern and the general apperarance cool gray. The interior designer made breakthroughs in expansion of the architectural appearance, but debates were caused due to the space positioning. the designer thinks the interior design serves the owner, with "people" at the center, making it comfortable and enjoyable. The style is just a presentation, even though it is very unusual, individual and cool. Therefore, the design gives up the single design style, which is in common use in American style and uses the special red color in China, which presents American-style atmospheric poise and also family warmth.

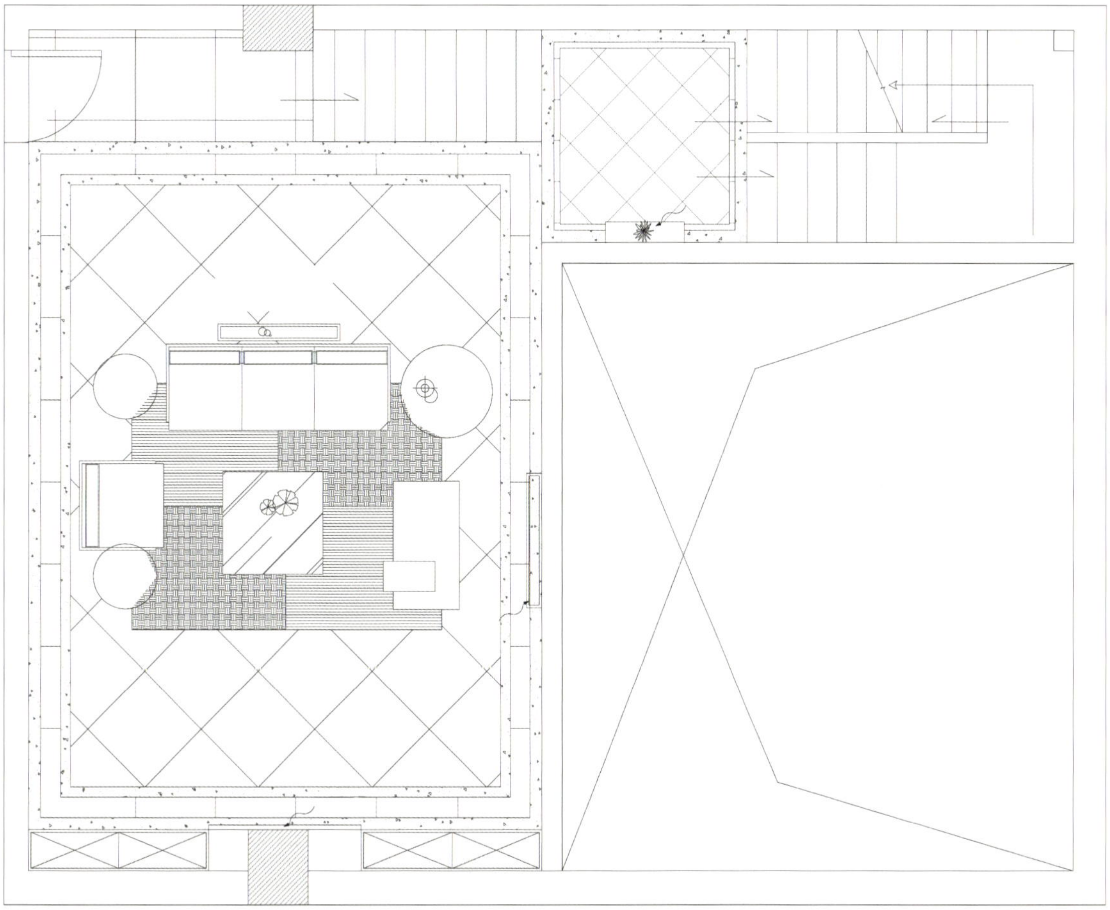

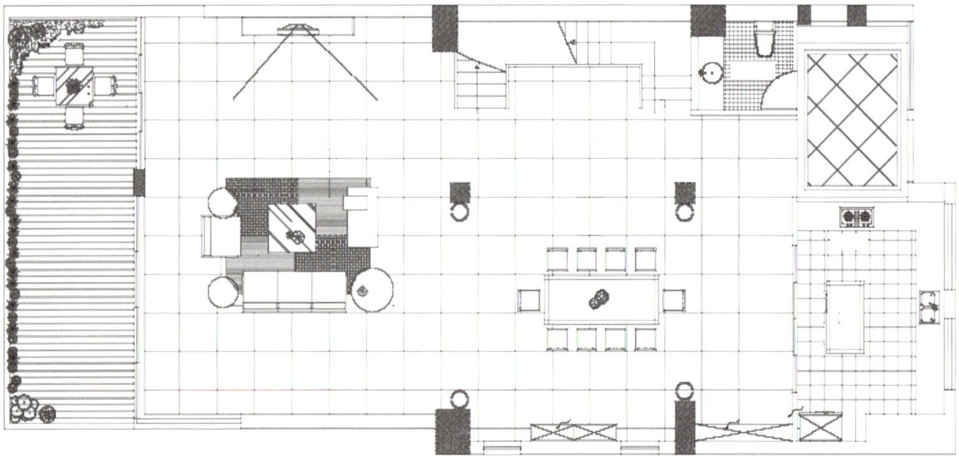

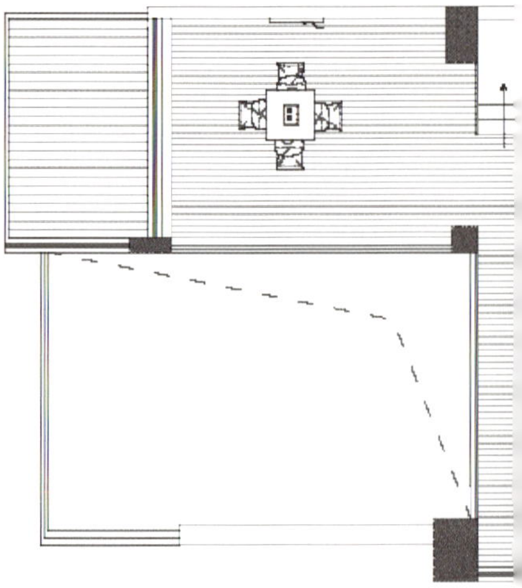

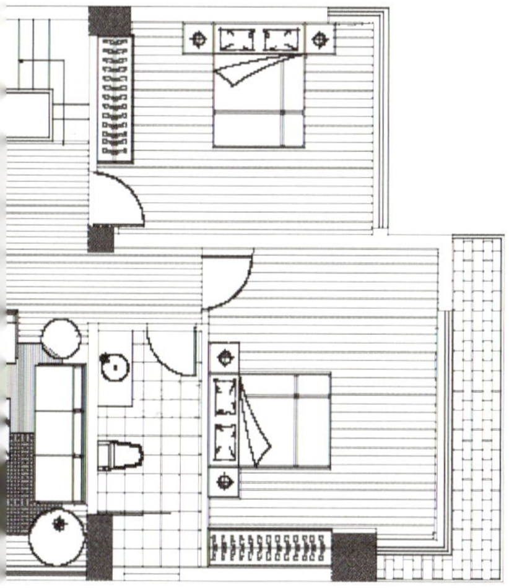

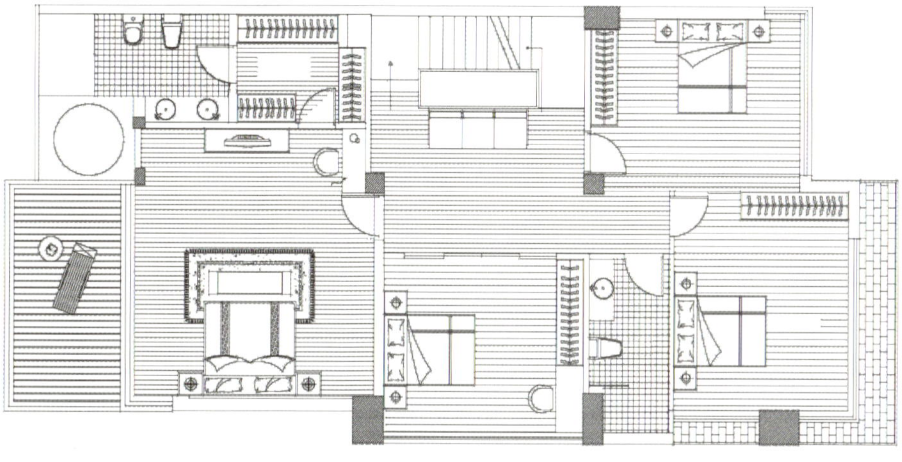

S-5 VILLA IN LGLY, BAOLI, HUNAN

湖南保利·麓谷林语 S-5 别墅

项目名称 湖南保利·麓谷林语 S-5 别墅 / 设计师 王小锋、王 赟 / 设计单位 广州尚逸装饰设计有限公司
空间面积 376 m² / 整体风格 新亚洲印象 / 投资商 保利地产

Project Name S-5 Villa in LGLY, Baoli, Hunan / Designers Wang Xiaofeng, Wang Yun / Design Company Guangzhou Sunny Decorative Design Co., Ltd / Area about 376 m² / Design Style New Asian impression / Investor Baoli Real Estate

本案设计不局限于形式上的奢华，而更趋向于一种各种材质之间的搭配运用，并着重于个性、时尚，户型各处无不流露着大自然的气息。以个性化的表达方式，体现超越现实的想象空间；用热情奔放的设计语言，描绘现代、个性、时尚的高品质生活。

Design of this case is not limited to superficial luxury, it more tends to the match between different materials, also emphasizes personality and fashion, every corner of the villa diffuses a kind of aura from nature. With a presentation of individuation, reflecting an imaginary space which is beyond reality; with the strongly passionate design language, successfully describing a modern, distinctive and fashionable life in high quality.

▶
在空间规划上，基本保持原建筑的功能布局。入口处和楼梯间的墙体隔断被打通，从而使入口处拥挤的空间顿时变得开阔。客厅沙发的布局换了个方向，使人一进来就能看见一面完整而大气的电视背景墙。原先客房的卫生间被开放出来作为公共卫生间，这样也解决了一层无公卫的尴尬。入口上空被利用起来作为二层的开放阅读室，不仅更加合理、最大化地利用了空间，而且也使二层和一层有了更好的沟通。三层的主卧分为卧室、衣帽间和主卫三个大空间，体现了当代东方人的居家生活理念。

As for the spatial structure, original function has mostly been kept. The wall between entrance hallway and staircase has been removed, the crowded hallway becomes much wider. Position of the furniture in the living room has been adjusted. Once you walk in, a splendid television background wall will first greet your eyes. The bathroom of the guest room now has been changed into a public one to solve the vacancy of the first floor. Upper space of the hallway is used as an open reading room of second floor, it appropriately makes the best of the space. Moreover, it connects these 2 floors. The master bedroom on the third floor can be divided into 3 parts: bedroom itself, coatroom and bathroom, that expressing life concepts of the contemporary orientals.

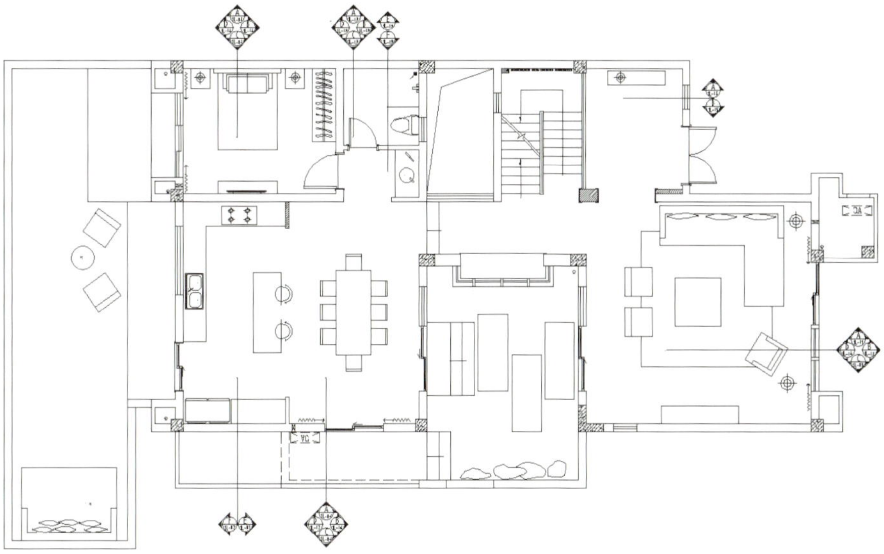

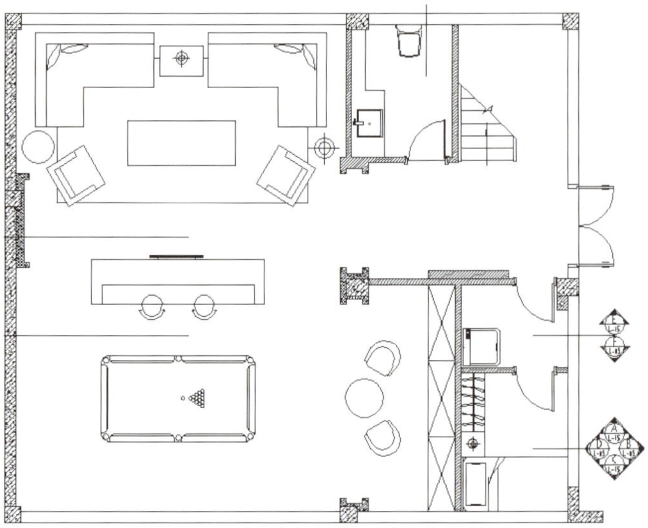

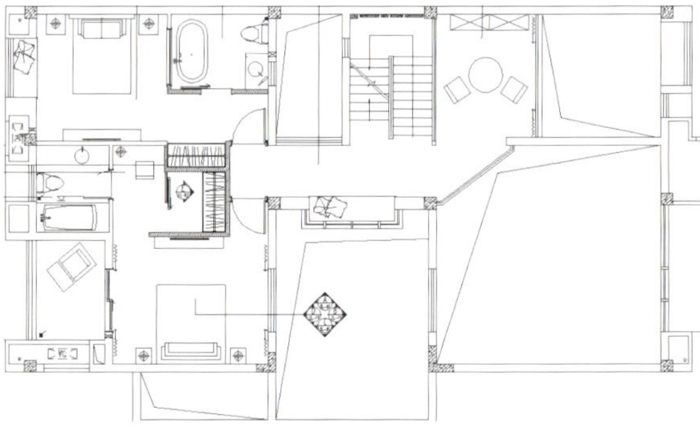

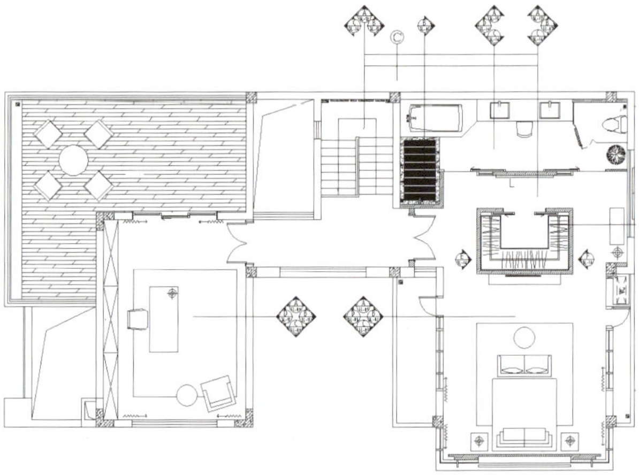

LINKOU VILLA
林口别墅

项目名称 林口别墅 / 设计师 郑闳泽 / 室内空间面积 350 m², 三层建筑 / 庭园面积 390 m²
设计主题 一楼巴洛克，二楼新古典低调奢华风

Project Name Linkou villa / Designer Zheng Hongze / Interior Area about 350 m² / Garden Area about 390 m²
Theme The Baroque On The First Floor, The New Classical Low-Key Luxury On The Second Floor

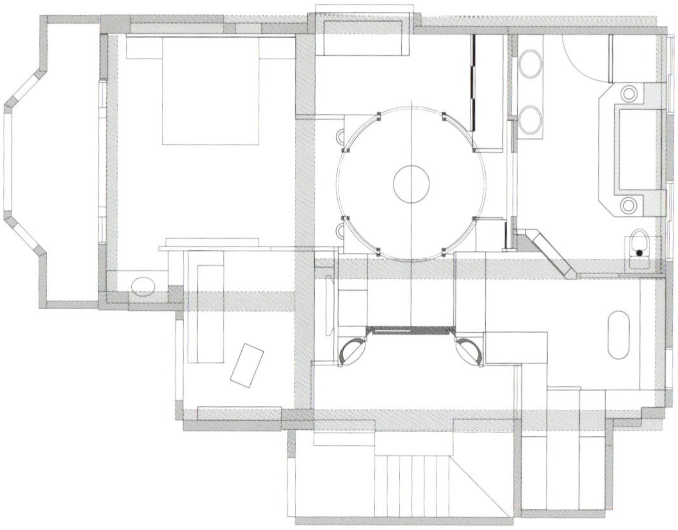

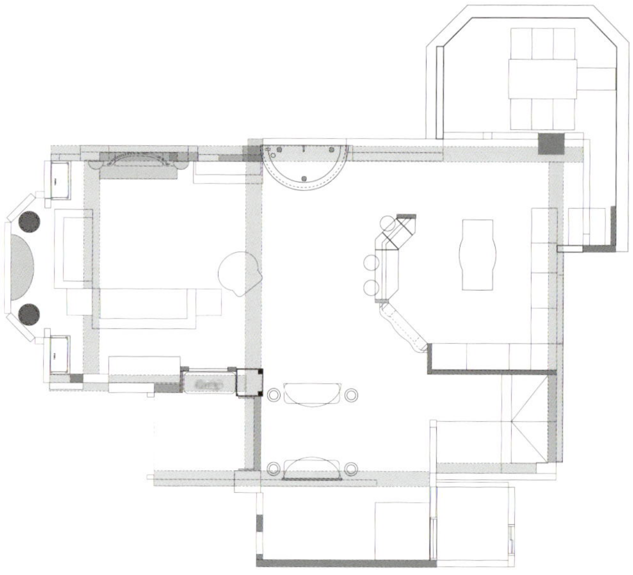

▶

巴洛克风格的客厅，搭配银狐大理石柱，饰面壁板均为开模定制品，地面采用帝诺大理石作无接缝处理，橱柜采用特殊开模定制美式橱柜。餐厅三面覆层铜条镶嵌玻璃。所有木作均作甲壳素涂布处理。主卧床头是价格高昂的雕刻定制品。浴室采用自动门装置。

A baroque living room with silver foxcolumnas espléndidas. Emperor's marbles of no juncture processing. Custom American-style cabinet. 3 walls of the kitchen covered with Copper glass. The main bedroom carved with a custom work valuing. The washing room with an automatic door device.

▶
儿童房，电脑室，储藏室——业主为年青企业家，经常往返欧洲参展，偏好德国样式，自地自建一座别墅，将年迈的母亲接来同住；又由于生活习性与饮食习惯均十分欧化，所以在设计上从形式到内涵均十分讲究。一楼属于男主人决定的空间，式样偏德奥的欧风样式，由于子女尚小，所以预留许多用于家庭派对、早餐以及节庆娱乐空间，并设计圣诞灯挂饰，以及情境遥控开关。二楼以主卧室为主，属于女主人决定的空间，同时由于高度限制，难以展现欧洲风格，所以设计上较为现代些。男女主人都喜欢穿着打扮，所以各有一间专属的更衣室，浴室也十分宽敞、舒适。三楼则是子女的活动空间、电脑室以及储藏室。

As a young entrepreneur, the owner always flies between China and Europe. He likes German-style architecture and builds a villa for his old mother. Since the life and dietary habits are very European, the design is very exquisite from the form to inside.
Due to the limits of the height, the European style can not be presented completeply and becomes modern. The masters like dressing up and each has a private changing room and the washing room is very comfortable and spacious. The third floor is activity space for children, including computer room and storeroom.

本色

顶级别墅空间 Top Villa Space

TRUE COLOURS

CONTEMPORARY AMERICAN
当代美式

浸润着优雅与简洁，沉淀着经典与朴素，
线条流畅、尺度适宜，
灰色阶色彩涂装，满满的和谐融洽，
关上耳朵，细细品味，淡淡的微笑爬上眉梢。

AMERICAN STYLE VILLA

美式别墅

项目名称 美式别墅 / 设计师 黄千祝 / 空间面积 270 m² / 主要建材 壁纸、实木地板、铁件、大理石、玄武岩+喷漆、玻璃、录音室专用吸音布板、鳄鱼皮料、水晶灯 / 项目地址 中国台湾桃园

Project Name American Style Villa / Designer Huang Qianzhu / Aera about 270 m² / Materials Wallpaper, Solid Wood Flooring, Iron, Maple Stone, Basalt + Spray Paint, Glass, The Studio Special Sound-Absorbing Cloth, The Crocodile Leather, Crystal Lamp
Location Taoyuan, Taiwan, China

业主希望打造一座梦想中的城堡，从买地开始逐步圆梦，建筑外观设计的灵感是从瑞士艺柏表博物馆而来，三层楼的独栋美式别墅设计。入口处的高高的柱子一字排开，搭配黑色地砖，让空间呈现宏伟、典雅的气势。入口处的造型屏风柜为整个公共区域的焦点，在巧妙的规划下，其正后方为客厅电视柜，侧面为餐厅电视柜体。客厅特意挑高设计，客厅主墙为喷金属漆的玄武岩。复古造型的家具搭配古典水晶吊灯，营造典雅华丽情境，并在餐厅旁以玻璃及不锈钢打造轻透的半圆形玻璃现代酒柜，同时在酒窖外门把上设计带有业主名字的家徽，从而增加独特美感。半弧形贯穿三层楼的旋转梯设计，以铁艺栏杆融合实木扶手创造细致的触感，7 m 长的施华洛世奇水晶吊灯，以简洁却奢华的空间语汇，如光影水瀑般宣泄而下，使楼梯犹如一件巨型艺术品。楼上为私人活动区，视听室壁板特别选用专业录音室内的扩散板，可以有效排除空间杂音，顶棚的吸声板为进口吸声壁布材质，以专业录音棚的规格、材质打造完美视听空间。

▶

The owner wants to have a dreamy castle. The inspiration of the facade comes from the Swiss Chipoba Museum and the house is American-style with 3 floors. The hallway is tall and the cylinders stand as a row, with black floor tiles, which give grand elegance. The screen ark is the center of the public space, with the rear the TV ark for the living room, the side as the TV ark for the dining room. The special tall design for the living room and the main wall with basalt and metallic painting, as well as the classical crystal droplamps bring a glorious and elegant atomasphere. The whirling stairs run though the whole house and the 7 meters long SWAROVSKI crystal droplamps hang like waterproof, which makes the stairs like a large work of art. The upper floor is private. The visual and studio room uses special diffuser plates which are usually used for professional studios, which can rule out noise efficiently.

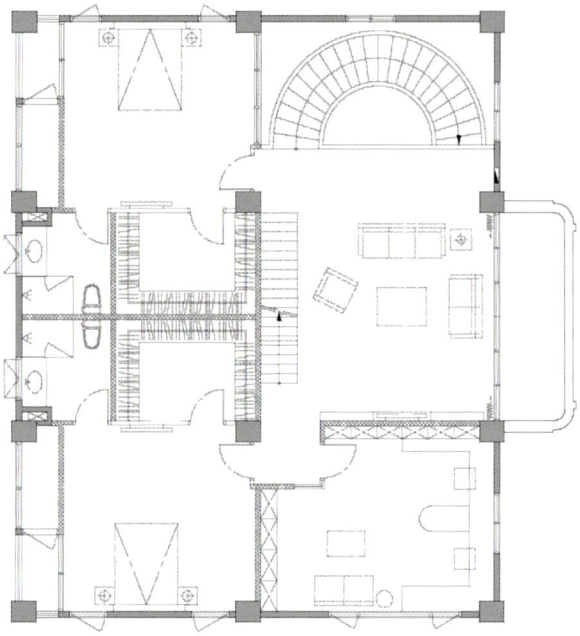

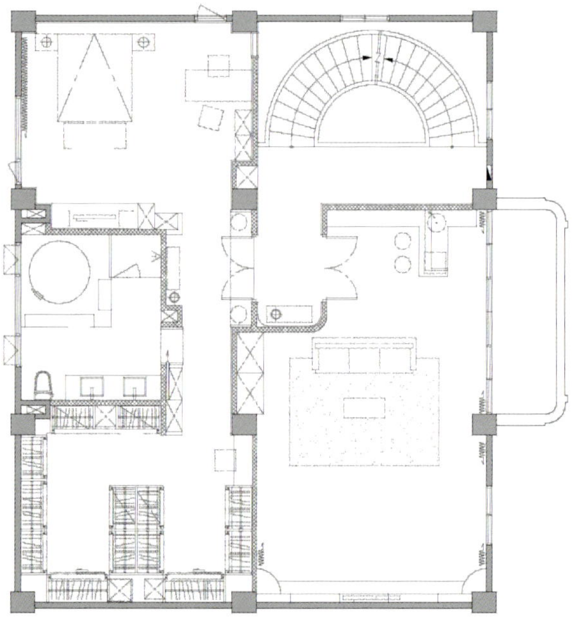

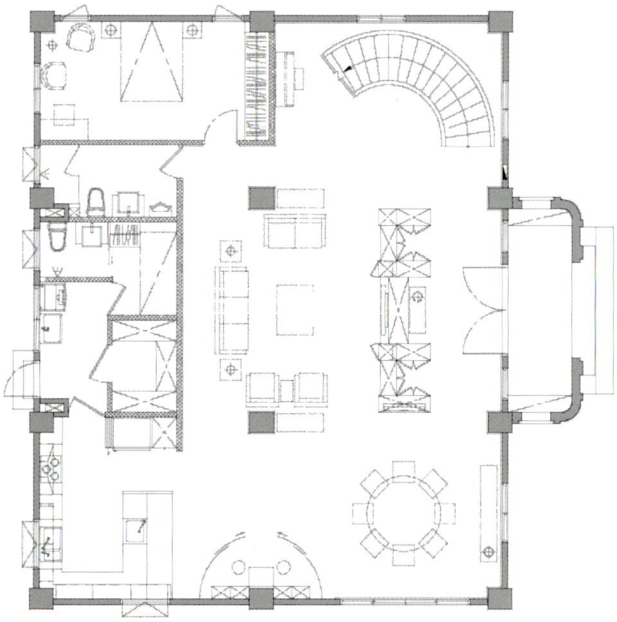

主卧室大红皮革床头背板，运用流行、新颖的材质以及大胆的色彩表现当代同时又带有复古的风格，并以白色为主色，预留空间让屋主装饰；主卧更衣室规划采用精品店的设计概念，柜内藏灯光及镜面增加了照明反射效果，兼具功能性及装饰性。
整体室内装潢及布置皆以新古典风格为主，并运用壁纸、刺绣、提花元素、线条华丽的古典家具以及各式水晶灯饰，共同烘托出宫廷般典雅、华丽的氛围。

In the master bedroom, the red backboard uses new materials. The bold color presents the modern and classical style. The changing room is just like a boutique, the clothes are arranged according to pure specification. The light and mirror hidden in the cabinets give reflection, intergrating the function and comfort.
With the new classic style as the theme, the interior designer uses wallpapers, embroidery, flower elements, classic furniture with luxuriant lines and crystal droplamps, which gives gorgeous and elegant atmosphere of a palace.

DOLOMITES HOUSE
多罗迈特山住宅

项目名称　多罗迈特山住宅 / 设计师　Jacopo Mascheroni / 设计单位　JM 建筑事务所
景观设计　Nippon Bonsai, Carate Brianza (MI) / 摄影师　Hermann Gasser, Jacopo Mascheroni

Project Name　Dolomites House / Designer　Jacopo Mascheroni / Design Company　JM Architecture
Landscape　Nippon Bonsai, Carate Brianza (MI) / Photographer　Hermann Gasser, Jacopo Mascheroni

这所住宅位于意大利和奥地利交界的地方，地处多罗迈特山。由于不舒适的住房情况，从而产生了重新整修的想法。目前，这个建筑有一个非常大而高的入口大厅，周围布满了房间，通常需通过大厅才能参观整个住宅。底层的厨房、餐厅、客厅与房屋的整个面积并不成比例，以至于所有房间的光线都不充足。

因此整个一楼都必须重新设计。经过改造后，一个用玻璃隔断的健身房隔开所有不同空间，几个新的房间设计在边缘区域，以此来丰富入口大厅和过大的家庭套房，包括厨房、休息室和学习区。

The idea of remodelling this house in the Dolomite mountains at the border between Italy and Austria started because of an uncomfortable living situation. The existing building presented a very large and high entry hall with the rooms arranged around it, and it was always necessary to walk through it to move around the house. On the ground floor, the kitchen, dining room, living room and the stube were not proportioned to the size of the house and they couldn't get enough lights.

For this reason, the entire ground floor has been demolished, and a new interior clear glass envelope with a gyp wall inserted is now the only separation among the different spaces. Few new volumes have been added around the house to extend the entry hall and the oversized family room, which include dining, relaxing and study areas.

▶
为了使各个房间获得更多的光线，定制的硅幕墙取代了房间的外墙，并设置放射形框和大型的滑门以增加室内和室外的互动。冬天温度很低时，通过地板反射、周边地板散热器、瓦斯和木质壁炉给房间加热。

▶
To allow the maximum amount of lights into the rooms, the façades have been replaced by a custom-designed structural silicon curtain wall, with base and head mullions flush with the floor and ceiling, and large pop-out sliding doors to create interaction between indoor and outdoor space. Because of the very low temperature of the winter, the heating solution is the combination of radiant floor heating, perimeter floor radiators along the curtain wall, a gas see-through-the-outside fireplace and a wood one.

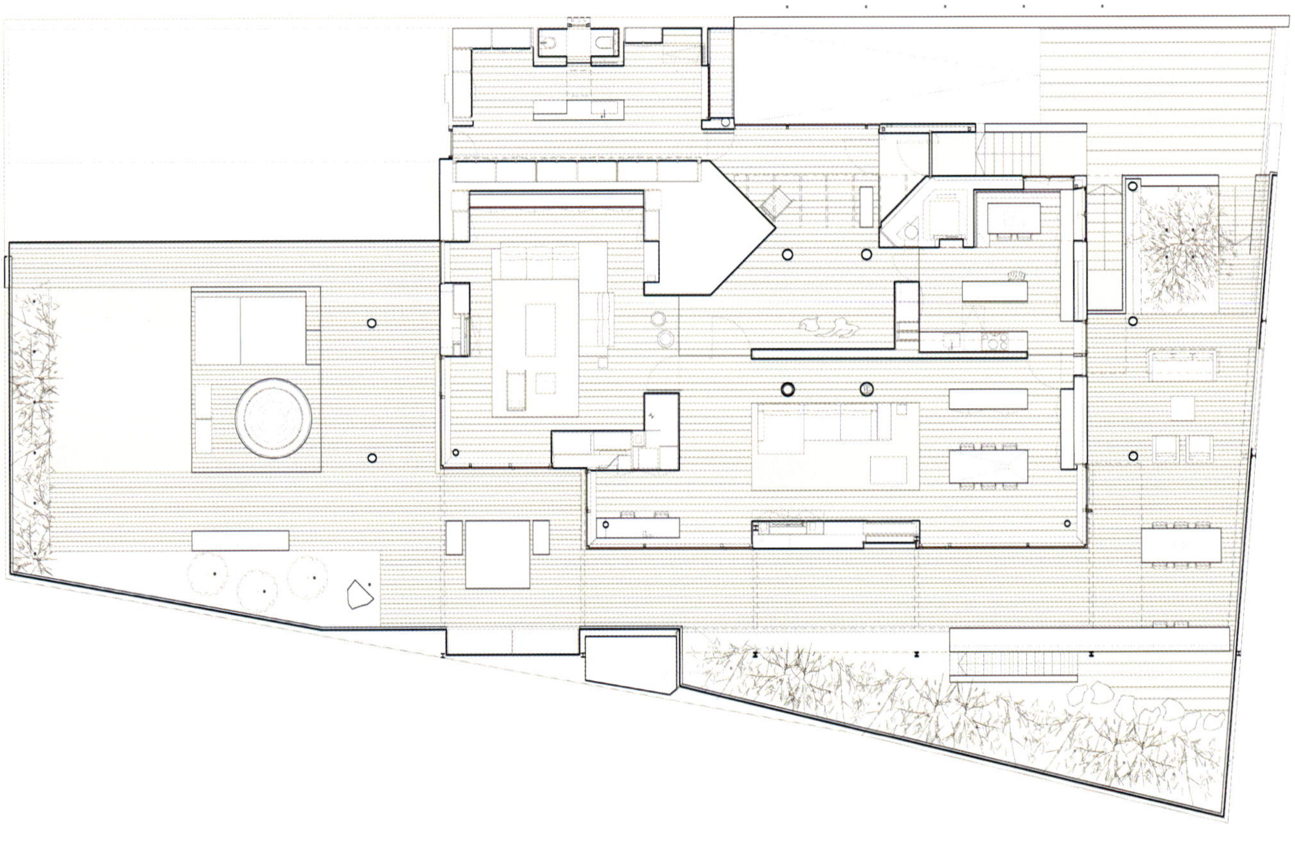

▶
整个住宅受指令总线系统完全控制，并特别备有音、视频设备包，括贴在墙上、顶棚以及其他地方的扬声器。几乎所有配置都是定制的，从整个厨房设备到桌子、凳子、瓦斯和木质壁炉。
花园被一道 2.5 m 高的木墙包围，设置木板加重光线的透射和投射效果。一个木质露台围绕着整个住宅，其上有一个日光浴台，配有一个按摩浴缸和两个长桌。
水平遮阳系统通过插入钢横梁而进行操作，使顶棚、室内地面地板在视觉上得到延伸。钢框架木伞使得在雨天业主也能坐在户外休息。

The house is controlled by a fully-automated iBus system, and particular attention has been paid to the audio-video equipment, which also includes invisible speakers plastered into walls or ceilings of different areas. Almost everything has been custom designed, from the entire kitchen to the tables, desks, gas and wood fireplaces.
The garden is enclosed by a 2.5m tall wood wall with horizontal slats to emphasize the perspective and cast shadows of sunlight from above and of spotlights from below. A wood deck surrounds the house, runs up a sunbathing platform with a Jacuzzi inserted, and covers two long desks.
The sun shading is a horizontal system that runs on tracks inserted in steel beams, as if the interior ground floor ceiling would be extended out. A steel framed wood canopy on one side allows to sit outside in a rainy day.

COMO LOFT

科莫阁楼

项目名称 科莫阁楼 / 设计师 Jacopo Mascheroni / 设计单位 [illegible] 事务所 / 摄影师 Jacopo [illegible]

Project Name Como Loft / Designer Jacopo Mascheroni / [illegible] Company JM Architecture [illegible] Jacopo Mascheroni

位于意大利科莫城附近的一座古老的修道院，需要整修并改造居住区，这样一来两栋相邻的住宅连在一起形成了一个两层公寓。通过对顶棚的不同方向和高度进行设计，建筑的外形有了很大的改善。设计的主要原则是保持整个外形不变，保留建筑原来的长度，尽量减小对外部空间的占用。具体措施就是嵌入一个封闭式盒子空间，一面包含一个小型的卧室和两个浴室，使住宅有一个具备主要功能的开放空间。主入口是一个很小的门厅，会给人一种压迫感。在走廊周围，客厅、厨房、餐厅用一个喷漆墙面与主人卧室分隔开来。不同顶棚间都有一个视频投影仪，在厨房墙面上播放画面。厨房墙面的柜子隐藏在墙面下，通过系统可以让它滑下来，这样做的目的就是最大限度的扩展投影放映面。另外一个特色是楼梯，由聚合钢梁做成。地板是精选的加拿大枫木，由于枫木在加热状况下不稳定，所以采取定制的方式在上面铺上 2.5mm 厚的加拿大枫木，底层加碎料板。

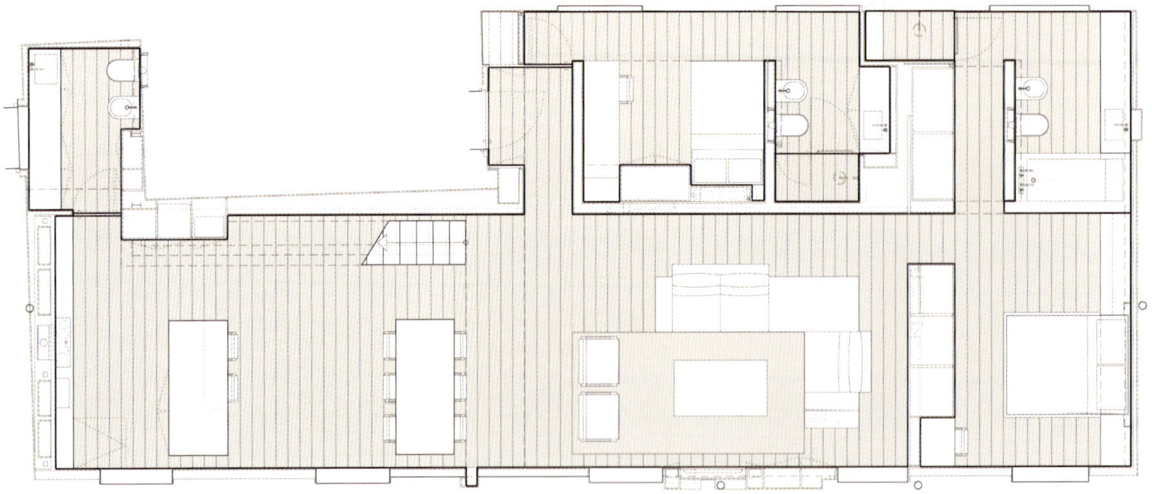

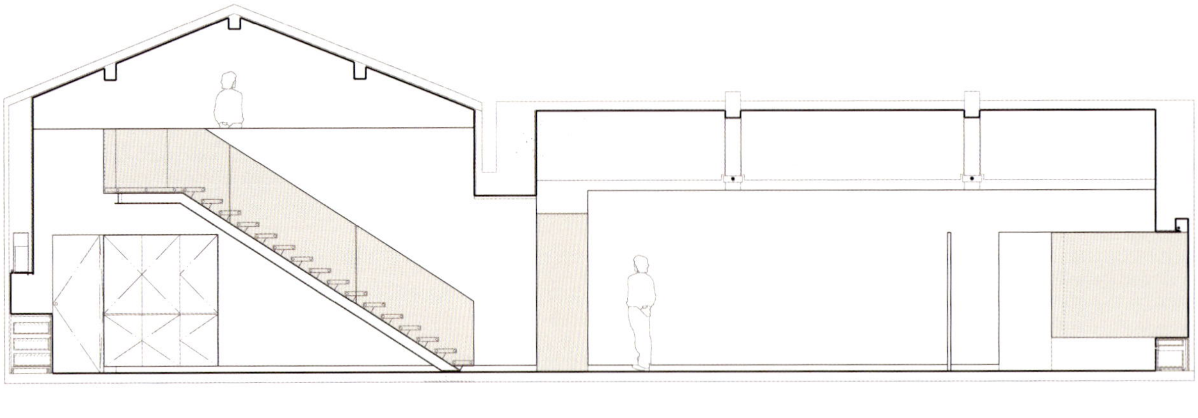

When an old monastery located around the city of Como, Italy, was being remodelled and converted into a residential complex, two adjacent units have been connected to form this duplex apartment. The shape of the original units is clearly enhanced by the different pitched ceiling direction as well as by their different heights. The main idea was to keep the shell intact, maintaining the entire length of the unit open and invading the space minimally. The result is the insertion of an enclosed box on one side containing a small bedroom and two bathrooms, which leaves an open space to the main functions of the home. The main entry is a very small foyer which creates a sense of compression before entering the main gallery, where the living spaces, kitchen, dining and living are separated from the master bedroom by a lacquered block. In the drop between the two different ceilings, a video projector is inserted to play images on the wall above the kitchen. To allow the maximum projection surface, the upper kitchen cabinets are hidden behind the wall and slide down with a motorized system. The other characteristic element in the main space is the staircase, which is an assembled steel beam with open treads made by a cantilever steel plate covered in wood. Canadian maple has been carefully selected for the hardwood floors, and since maple wood isn't stable with radiant floor heating, the floor planks were custom built with 2.5mm Canadian maple top and back layers on a particle board. The 20cm width of the floor planks modulate the position of all the elements, from the walls to the millwork. The floor planks also run up the master suite wall, which is lit by an upper light cove.

所有周边墙壁和顶棚均隐藏了额外的保温层，浴室的顶棚还配备了辐射板。相邻的墙已覆盖了 LED 层，提供额外的隔声装置。这房子内所有配置都是完全自动化的，并且是由位于二层的服务器控制。所有的家具都是定制的，包括整个厨房、浴室水槽、床、餐桌、桌子和浴池。

All perimeter walls and ceilings hide an additional layer of thermal insulation and the bathrooms are equipped with radiant panels in the ceiling. The walls adjacent to other units have been covered with a LED layer to give additional acoustical insulation. Glacier white Corian is used for counter tops and sinks in the kitchen, laundry room and bathrooms, for the bath-tub, showers and parts of the bathroom walls. The house is fully automated and is managed with a server located in the studio on the second level. Everything has been custom designed, including the entire kitchen, bathroom sinks, beds, dining table, desks and bath-tub.

ADRIA 2-BEDROOM SHOWFLAT

阿德里亚两房样板房

项目名称 阿德里亚两房样板房 / 设计单位 ONG&ONG Pte Ltd. / 项目地址 新加坡达比莎路 21 号
客户 Jumbo Valley Pte Ltd. / 摄影师 See Chee Keong

Project Name Adria 2-Bedroom Showflat / Design Company ONG&ONG Pte Ltd.
Location 21 Derbyshire Road, Singapore / Client Jumbo Valley Pte Ltd. / Photographer See Chee Keong

阿德里亚两房样板房距离乌节路仅仅数分钟路程，是时尚都市生活的象征。独具品位的单色调在每一道灰墙上展现出空间感，并与白色爵士白大理石地板形成对比。简洁的流线营造出美妙的视角，某些位置别出心裁地设置了镜子，从而增强了空间的视觉深度。

Mere minutes away from Orchard Road, Adria's 2-bedroom apartments embody the sophisticated modernity of city life. Monochromatic tones stylishly present each space within dark walls, contrasted against white Volakas marble floor. Simple linear elements create interesting perspectives, while mirrors are cleverly employed to enhance the apartment's depth.

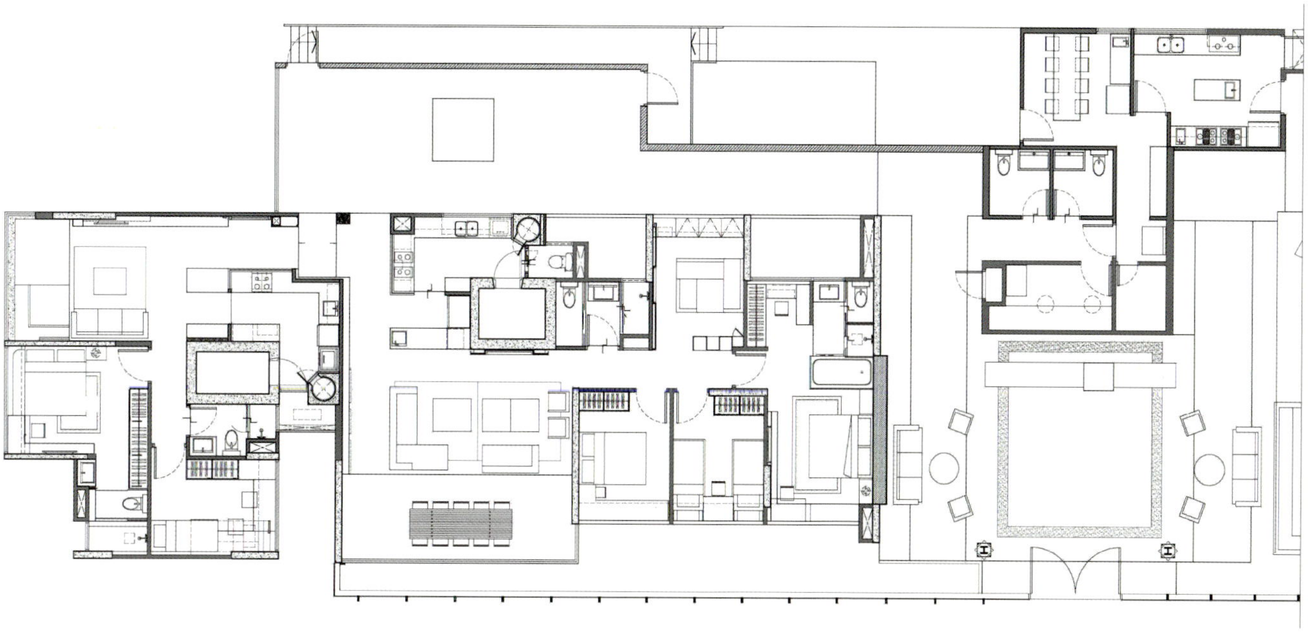

POSSIBLY MAYBE

本案中的厨房、客厅以及用餐区贯穿为一体，精致而奢华，开放式的空间格局使住宅显得更加大气。厨房的设施齐全，简洁而优雅的白色橱柜提供了干净、时尚的烹饪平台。
儿童房中构造新颖的双人床与玻璃书桌散发出奇特的味道，而主卧以一面视觉效果强烈的石墙为亮点。这栋住宅无疑将简洁风格表现到了极致。

Communal areas are compact yet luxurious with the fusion of kitchen, living and dining areas. This seamless transition between areas gives the apartment an impression of greater space. The kitchen is well appointed with simple yet elegant white cabinets that create a clean, modern canvas.
An interestingly layered bunk bed and glass study desk give the children's room an element of quirkiness, while a dramatic stone wall in the master bedroom serves as a key visual element. These apartments demonstrate simplicity at its best.

PENTHOUSE DOWNTOWN MONTREAL
蒙特利尔商区阁楼

项目名称　蒙特利尔商区阁楼 / 设计师　Martine Doyon
设计单位　René Desjardinsrené Desjardins

Project Name　Penthouse Downtown Montreal / Designer　Martine Doyon
Design Company　René Desjardinsrené Desjardins

业主之前在蒙特利尔姊妹岛区的大房子里住了很多年，现在选中了在蒙特利尔市中心区蒙特利尔皇家山和圣劳伦斯河这里的一个全景公寓。一开始，我觉得这一全蒙特利尔市最棒的壮丽美景应该是项目的设计焦点。业主想要一套明亮、温馨的洛夫特风格的公寓。设计师建议打造一个像是曼哈顿或者洛杉矶的当代大型洛夫特空间，安静、舒适，同时也可以看到城市繁复快速的变化。

The client, after spending many years in a large home on Ile-des-Sœurs, had taken an apartment with a panoramic view of downtown Montreal, Mount Royal and the St. Lawrence River. From the start, it seemed to me that this magnificent view – one of the best in Montreal – should become the focus of the project. The client wanted to live in a bright, warm loft-style apartment. I suggested a space much like the large contemporary lofts of Manhattan or Los Angeles, a loft that would speak of calmness and comfort yet look out over the continually changing showcase of the city.

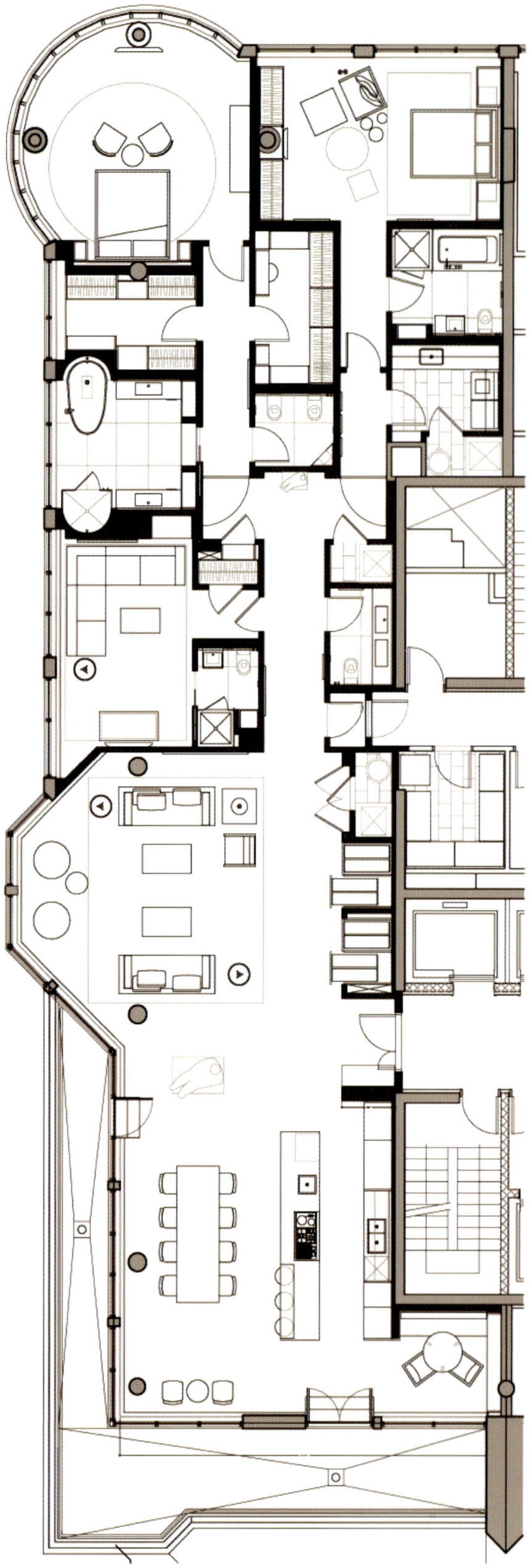

因为大楼还在建设中，一切都有可能。这套公寓在 23 层，包括两个阁楼，面积一共为 3,300 m²。作为一个出发点，这里除了是通往其中一个阁楼的走廊现在也成了房子的电气系统区。这就意味着公寓可以是纯矩形，横跨楼高，也就是北、东、西边都可以设计全高窗，用来接收自然光。

该公寓由两个区组成：一边是公共区，包括起居室、餐厅、厨房和家庭影院间；另一边是业主自己的和孩子的私人房间。在色彩装饰上，设计师利用白色、灰色和棕色打造了一个温馨、简约的空间环境。

Since the building was still under construction, everything was possible. The space consisted of a large 3,300 sq. ft. unit assembled from two penthouses on the 23rd floor. As a point of departure, space intended for a corridor to one of the penthouses could now be used to house the mechanical and electrical systems. This meant that the apartment could take the form of a great pure rectangle spanning across the full length of the building, with full-height windows on the North, East and West sides that would provide exceptional natural lights.
The apartment comprises two zones: on one side, the shared areas, the living room, dining room and kitchen as well as a living area/ home theatre; and on the other, the private quarters of the client and his son. Detailed work began on the interior envelope, playing with different coatings and colours for the volumes, using whites and shades of grey and brown to create a pared-down, warm environment.

PRIVATE APARTMENT-NEW WESTGATE GARDEN-SHANGHAI

私人公寓 – 老西门新苑 – 上海

项目名称　私人公寓 – 老西门新苑 – 上海 / 设计师　Thomas DARIEL & Benoit ARFEUILLERE / 设计单位　Lime 388 by Dariel and Arfeuillere
空间面积　150 m² / 项目地址　中国上海 / 客户　Jumbo Valley Pte Ltd. / 摄影师　Tristan Chapuis

Project Name　Private Apartment - New Westgate Garden - Shanghai / Designers　Thomas DARIEL & Benoit ARFEUILLERE
Design Company　Lime 388 by Dariel and Arfeuillere / Area　about 150 m² / Location　Shanghai, China
Client　Jumbo Valley Pte Ltd. / Photographer　Tristan Chapuis

本项目是在上海市卢湾区西藏南路的高端住宅楼第 29 层的一个 150 m^2 的私人公寓。公寓位于在上海繁华的中心区，庆幸的是其位于出名的娱乐潮流中心新天地附近却能保持相对安静的氛围。

设计师 Lime388 认为对于这个公寓的设计要融合个性和实用性功能房个人的和专业的以及客户的和时尚设计师的需求。因此空间规划了基本的功能房——卧室、厨房、起居室、更衣室，并且这些功能房围绕中间的工作区，即圆形的玻璃区域，既突出了空间感，又显示出了设计的创意。柔和的曲线传递着优雅和诗意，黑白条纹体现出一种生动活泼的波普风。

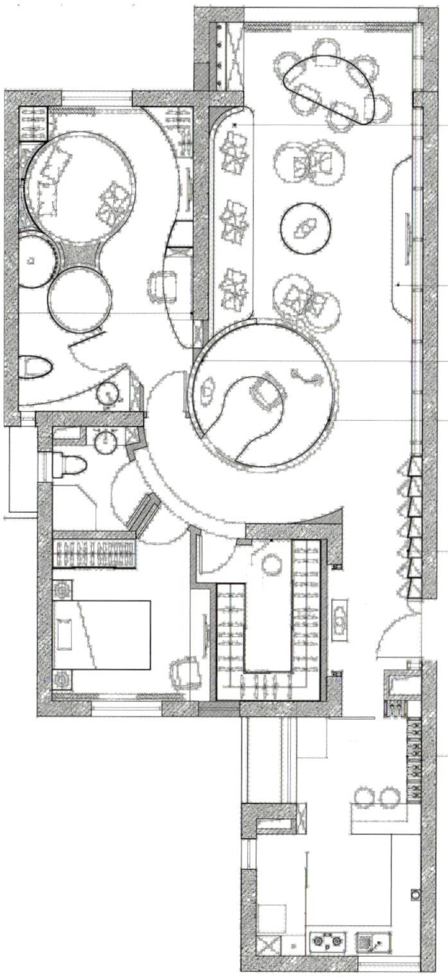

This project is a 150m^{2}private apartment located on the 29th floor of a high-end residential building on South Xizang Road in Luwan District. In the heart of Shanghai bustling downtown, the place benefits from the proximity of the well-known entertainment and lifestyle center that is Xintiandi while staying a quiet nest.

For this apartment, the designers of Lime 388 had to consider the personality and the needs, personal as well as professional, of their client, Fashion Designer. The space layout is therefore composed of basic functional rooms – bedrooms, kitchen, living room, changing room, which are yet all organized around a central working area, glass round shaped, that highlights the space and reminds the creativity of its inhabitants. Curves and soft lines convey elegance and poesy while large black and white stripes give a vivid POP feeling.

140 DETIGER LOFT

140 Detiger 公寓

项目名称 140 Detiger 公寓 / 设计师 Erin Sonntag & Markus Dochantschi / 设计单位 StudioMDA
空间面积 396 m^2 / 项目地址 美国纽约曼哈顿

Project Name 140 Detiger Loft / Designers Erin Sonntag & Markus Dochantschi / Design Company studioMDA
Area about 396 m^2 / Location Manhattan, NY, USA

▶
此套公寓在设计上与其他曼哈顿北区的周边邻里有些对立，唯一的共同点就是采用了开放式格局设计。此项目的设计主旨在于创造出画廊式的居住空间，同时将休息空间独立出来，赋予空间独特性和私密性。
儿童墙让儿童有钻进家中小窝的愉悦感，一系列玩具橱柜、人工草皮铺设的游戏区、阁楼式双层床、舒适温馨的茧式躺座都充满了活力和趣味性。茧式躺座传达着它的寓意，为具有想象力的想法的孕育提供避风港，在这里，人们将沉浸于无穷无尽的想象之中，或享受阅读的乐趣。

TAMA

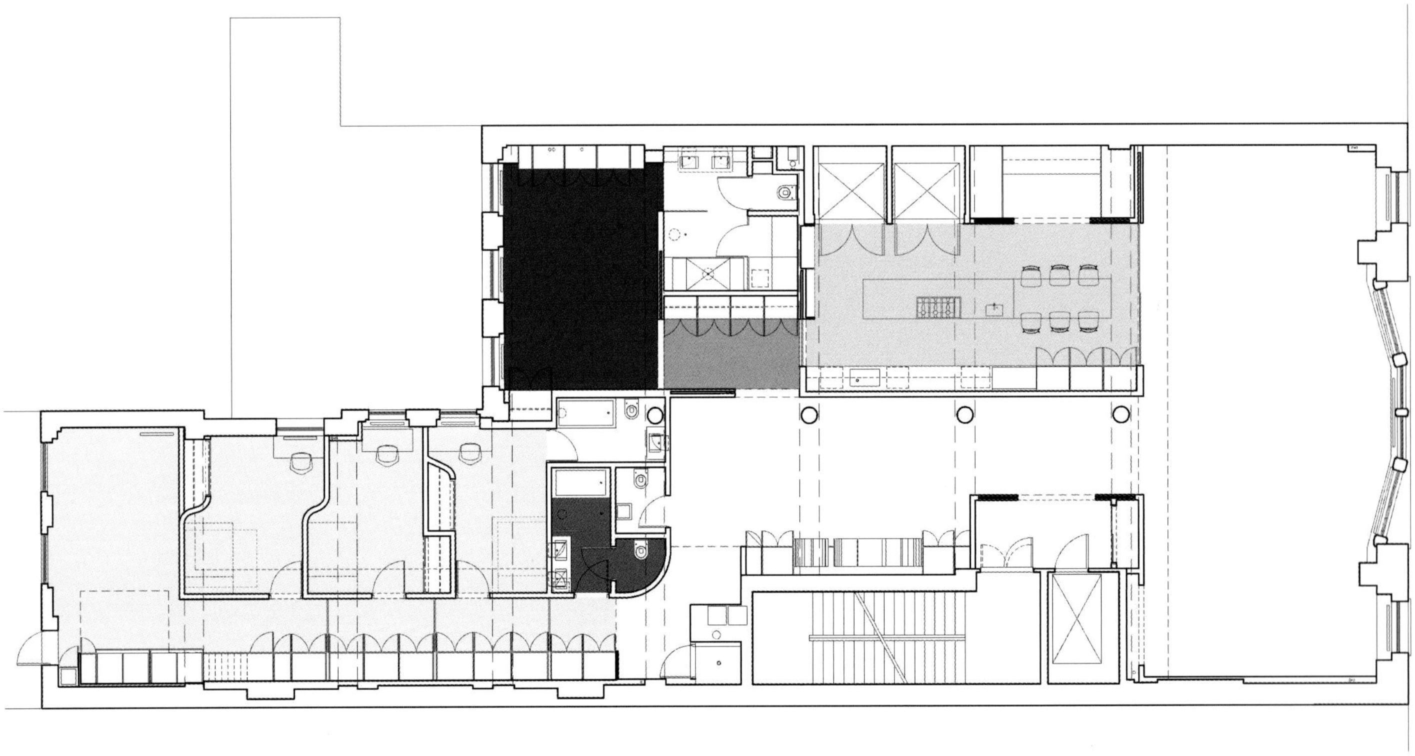

▶
The design of the loft is in a way antithetical to Manhattan's Noho neighborhood, where the common practice is the open plan. The approach for this residence was to curate gallery-like vistas in the living areas and separate the sleeping spaces to allow each its own identity and privacy.
The children's wall provides a home within the home, combining storage for toys, an Astroturf covered the play area, a lofted bunk, and nestled, intimate "cocoons" for leisure and reading. Situated to the scale of a child's hiding place, the wall encourages adventures in the labyrinthine corridors of imaginations. The cocoons endeavor to achieve the implied metamorphosis, by providing sanctuary for imaginative thoughts, where one may lose oneself in daydreams, or in some great books.

LOVE

RESIDENCE IN MEXICO
墨西哥住宅

项目名称　墨西哥住宅 / 设计师　Pascal Arquitectos,George Cameron Nash,Baker Knapp & Tubbs Hajj Design,Christian Liagre
项目地址　墨西哥城 / 摄影师　Jaime Navarro

Project Name　Residence in Mexico / Designers　Pascal Arquitectos,George Cameron Nash,Baker Knapp & Tubbs Hajj Design,Christian Liagre / Location　El Pedregal, México City / Photographer　Jaime Navarro

这个项目的灵感来自于现当代奢华风格。宽阔敞亮的休闲娱乐空间，自然材质也显得更加舒适、温馨。房子是两层的，一层为公共空间区，还包括厨房等家居生活空间，二层有三个卧室、洗手间和电视房。这个项目依照家庭旅馆样式设计，有竞技房、酒吧、酒窖等，是个可以容下不少人的家庭房。

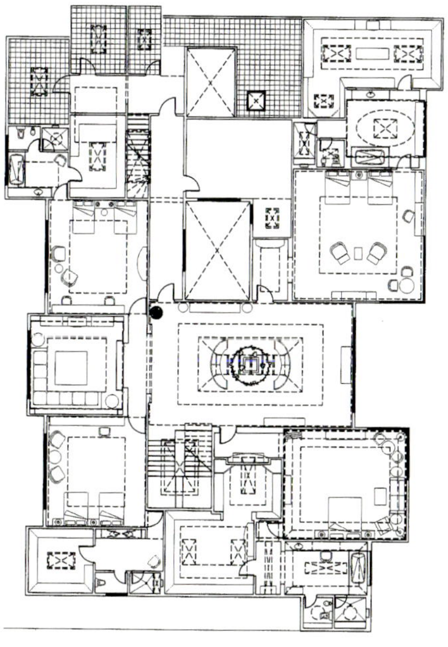

This project is inspired in the luxury and spaciousness of large villas expressed in a contemporary or modern style. Leisure and recreation areas spite being vast, are comfortable and warm due to the use of natural materials. The house is designed in two levels, the main floor houses public areas, kitchen and domestic services and on the upper level three bedrooms, restrooms and a TV room are located. The design concept in this project revolves around foyer hotel style areas, that welcomes to discover the trophy hunting room, the bar and the wine cellar and a family room that can accommodate a large concurrence.

FANTASY ISLAND
梦幻之岛

项目名称 梦幻之岛 / 设计师 Miroshkin Michael, Miroshkina Elen (Geometrix Design) / 空间面积 130 m²
项目地址 俄罗斯莫斯科 / 摄影师 Zinur Razutdinov

Project Name Fantasy Island / Designers Miroshkin Michael, Miroshkina Elen (Geometrix Design)
Area about 130 m² / Location Moscow, Russia / Photographs Zinur Razutdinov

我们旨在设计舒适的现代室内空间，将美学与便捷性及功能性集为一体的空间。
此项目的基调在策划期就已成形，我们使用“运动向量”计算，在公寓里各区域之间有序不乱地设置各种独特的墙面。

We've set a goal to organize the light modern interior which combine convenience, aesthetic qualities and functionality of space. The foundation of delivered tasks was chosen as far back to the planning stage. We've set up the direction, "the movement vector" which interlay the way logically through all the apartment's zones, with the help of interesting wall architecture.

POSSIBLY MAYBE
POSSIBLY MAYBE
SENSATION OF TIME
SENSATION OF TIME

▶

通过入口走廊，进入客厅，玻璃隔断将客厅和厨房分隔开来。这个设计合理地运用在每个空间，同时让空间显得紧凑，也能让它变得独具一格，明亮且通风。室内设计的重点是家庭影院区，这不仅是客厅的必要组成部分，而且也是体现艺术风格的展示区。电视面板可以移动，内嵌在黑色玻璃面墙体内，周围设置 LED。这种结构设置效果很好。这种情况下，美学主题并不会与功能性相冲突。在玻璃墙下边有衣柜抽屉。这种设计风格一直延续到顶棚，亮丽的面板内置有灯具，根据心情可以选择室内灯光的颜色。

装饰的重点在高光泽黑色大门，这些大门延伸了顶棚的视觉高度，同时黑色的魅力并没有在延伸中弱化。这种对比得以完美地体现，不仅在于灯光的选择，而且也决定于使用的材料。平滑、光亮的主体结构同舒适的木质形成对比，但是却异常和谐。两种材料在相互映衬下更加展露出各自的魅力。木质的运用使整个室内空间变得更加完整和统一。

厨房有个很特殊的东西，即带远程控制的桌子。只要按一下控制按钮，就能得到餐桌或者酒吧柜台。

卧室有很多大窗户，并设有独立的卫浴室和穿衣间。采用磨砂玻璃作为隔间，将自然光引入卫浴室。同时，天然的原料，如木头和石头也被应用，这两种材质的美在光线的组合下更加凸显。

梦幻和现实穿梭在整个项目中，它们演示了华丽的关于美学和功能的“二重唱”。

Through the entrance lobby we get into the large sitting-room, where with the help of extendable glass partitions the kitchen area is detached. This decision determines the prescription of every zone logically, but at the same time doesn't overload the space and makes it unique, light and airy.

The main emphasis in the interior is focused on the home theater zone. This is not only the necessary part of the sitting-room but also the complete art-object. The switched off TV panel "ad notam" becomes the invisible part of a big black glass panel with the all-round embedded light-emitting diodes (LEDs). The construction arrests sight and its prescription astonishes. And in this case the idea aesthetics doesn't pinch the functional purpose: we mean, the space for keeping CD and books on each side. Under the glass there is the commode with drawers, and all the construction continued straight to the ceiling in the form of lustrous panels with the built-in luminaires (lamps). The luminous decisions wide choice makes the interior change according to mood preferences.

The important decoration elements are high black lustrous doors. With their helps we enlarge the ceiling level and preserve the black colour. The contrasts are observed not only with the help of colour decisions but also with the materials which were used. The primary structure of smooth lustrous contrasts harmonically with homy (cozy) wood. Both of the materials underline the advantages of each other very profitably. Wood is also the "connecting" element which makes the interior integral and complete. The panels in the entrance lobby run smoothly to the kitchen zone, where that panels are emphasized with the massive wood board on the floor.

There is the "jack"(special thing) in the kitchen: the transformer –table with the remote control. One touch-and we get the dining-table, or bar counter.

The bedroom with a big quantity of windows has the separate WC-and-bathroom unit and the dressing room, of course. We've taken in (admited) the natural light into the WC-and-bathroom unit with the help of frosted glass, which was used as a partition. Also there were used natural materials as wood and stone. The beauty of two noble materials was underlined with the combination of colours.

Fantasy and reality were interlaced in this project. And they made the wonderful "duo" of aesthetics and functionality.

CASA Q
卡萨 Q

项目名称　卡萨 Q / 设计单位　Longhi 建筑事务所 – Luis Longhi
空间面积　420 m² / 摄影师　Juan Solano, CHOlon Photography

Project Name　Casa Q / Design Company　Longhi Architects – Luis Longhi
Area　about 420 m² / Photographer　Juan Solano, CHOlon Photography

一望无际的沙丘从沙漠向东往太平洋海岸悬崖翻滚而去，卡萨 Q 西面的渔民在此海岸经常出没。独特的自然地理环境使卡萨 Q 成为建造于秘鲁首都利马以南 117 km 未被占领的神秘海滩俱乐部的第一居所。

Infinite rolling dunes from the desert to the East and rocky Pacific Ocean cliffs used by fishermen to the West converge on the site of Casa Q; creating a unique natural environment. Casa Q is the first residence built in one of the areas not yet occupied at the Beach Club Misterio located 117 km south of Lima, Peru.

▶

卡萨 Q 打破环境的寂静，实现一对年轻夫妇“漂浮空间”的梦想，象征他们对未来家庭的展望。此空间用圆柱体支撑，自然地交叉分布而非强迫性地以网状分布。柱体呼应着住宅公共区域的玻璃推拉门；根据社交场合或天气情况，可打开或关闭玻璃面板，将露台作为起居或用餐区域单独使用，或将两种功能集合使用。
房子的其他区域，如客房、厨房及设施房，则在倾斜场地的背后隔离出来，并被赋予视觉上的美感，与前部修建的供娱乐、游泳使用的水库形成对比。宅中的任何一个视角都能瞭望无边的天际线。

▶

Challenging the stillness of the surroundings, Casa Q materializes the dreams of a young couple in a "floating volume" which embodies the spaces for a future family. The volume is supported by circular columns placed by intuition, as a dance, instead of forcibly in a grid. The dancing columns are accompanied by sliding glass panels that define the common area of the house; living-dining and terrace are integrated or separated by the option to open or close the glass panels depending on social and weather conditions.

The rest of the house-guest rooms, kitchen and services-are tastefully secluded at the back of the sloped site thus providing visual contrast with a volume of water in the front which has been unearthed for the enjoyment of swimming. Each view in the house connects to the infinite of the horizon.

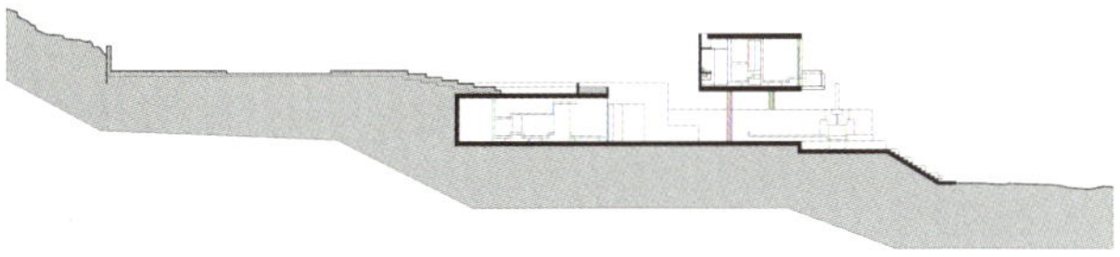

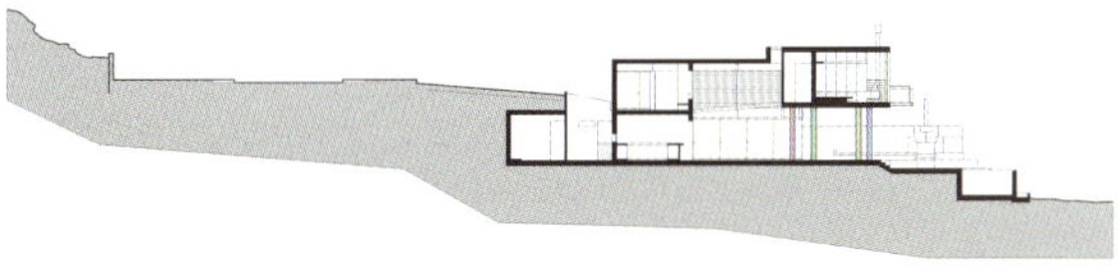

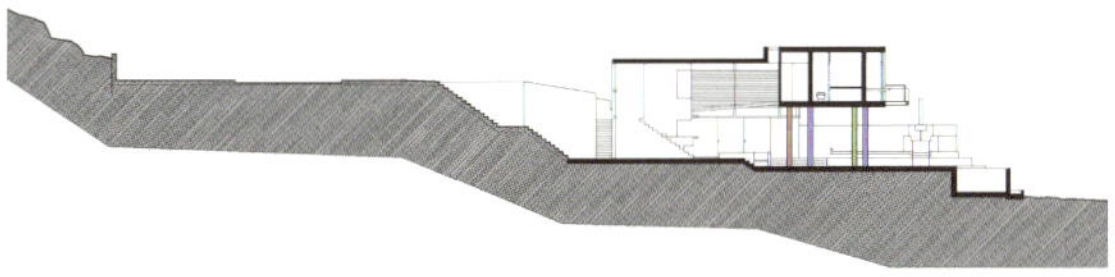

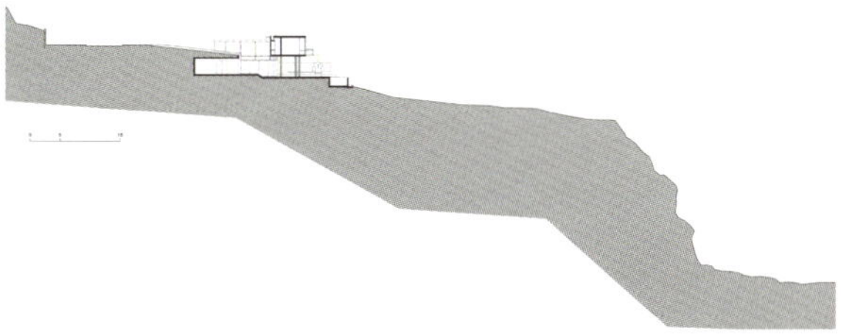

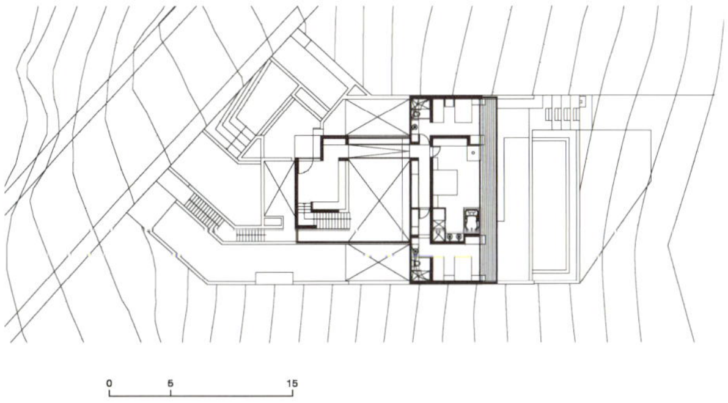
0
5
15

KATAMEYA HEIGHTS RESIDENCE

卡塔米亚山庄别馆

项目名称 卡塔米亚山庄别馆 / 设计单位 Mohamed Fares, 炼金术设计工作室
项目地址 埃及开罗 / 摄影师 Hussein Shaaban

Project Name Katameya Heights Residence / Design Company Mohamed Fares, Alchemy Design Studio
Location Katameya Heights, Cairo, Egypt / Photographer Hussein Shaaban

位于卡塔米亚山庄的这个别馆在众多的同类房屋中脱颖而出。无论是建筑还是室内设计都因各种可变换的特性而显得独特和出众。
设计师从项目的外观设计上着手，对房屋进行化茧为蝶的过渡。他比较注重设计的若干细节；颜色略有改变，用中性的灰色调与其他物件和谐搭配，显著地改善了外观整体氛围。用板岩贴饰部分墙面和现代设计的窗户边。在整座现代感极强的室内空间里加进了多种建筑设计元素，包括将部分的贴砖用木条替换。

Located in Katameya Heights, this residence stands out amongst its fellow homes in more ways than one, as the architecture and interior design are made unique by their altered nature
Alchemy's Mohamed Fares began his transition from ordinary to extraordinary with the exterior of the house. He accents the basic design with several details; the colours are slightly altered with a neutral grey tone and while this produces a significant effect on the overall mood, it remains in harmony with the rest of the compound. Slate tiles cover portions of the facade, playing a role in the more contemporary window dressings that were also adjusted. Several other elements add to the contemporary feel of the architecture including a shift away from brick complements to wooden strips and linear accents instead.

▶
为了更好地满足业主的生活方式和喜好，室内分区进行了重新设计，同时，柱子和板块的运用也使室内显得优雅和从容。本案的设计理念是打造一个充分融入个人生活，同时又能很好地保护隐私的生活空间。项目运用了现代的极简主义手法和充满诗意的设计风格。

Within the house more changes occur. The interior division was restructured to better suit the clients' lifestyle and columns and slabs were used for a more elegant outcome. The main concept was to create a space that integrates living areas but also maintains privacy. The design approach is a contemporary one with minimalistic tendencies and a poetic flair.

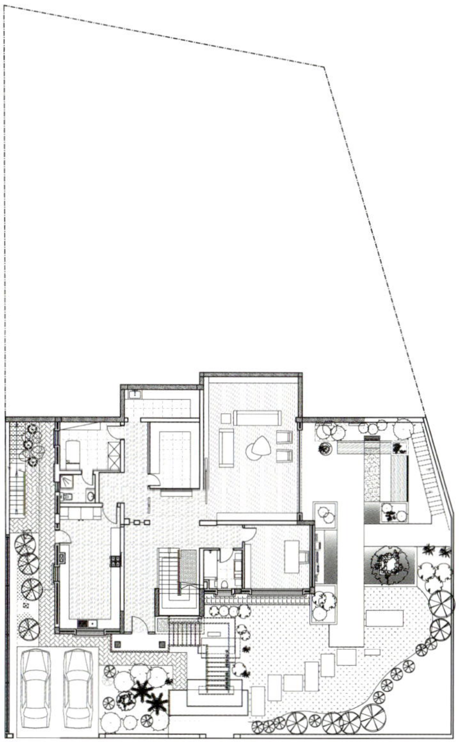

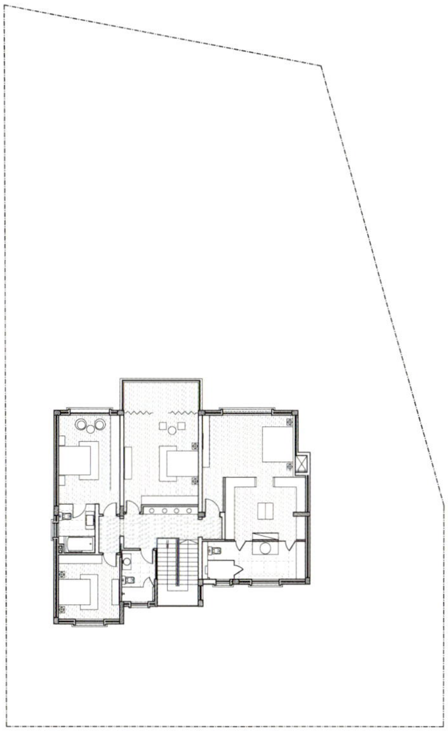

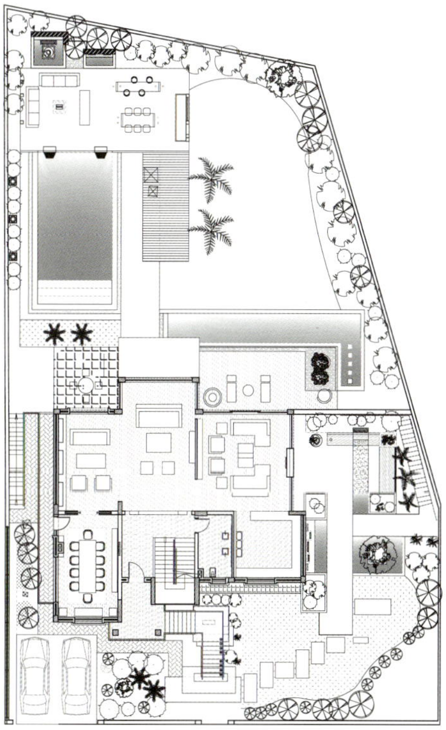

设计中有几点关键的地方，其一，并且最为令人印象深刻的是楼梯的设计。这一部分的关键之处并不全是因其设计创意的独特，更重要的是它对房子整体设计所产生的影响。撤掉原有的普通楼梯，并在原有楼梯处地面对应的顶棚上安装石榴蓝吊灯。撤掉了功能性楼梯，对于装修工人来说也是一个巨大的挑战。
厨房是另一个设计重点。采用了波菲风格的极简主义设计来体现高科技厨房的现代感。厨房用品和家电隐藏在简洁线条的柜体中，整体厨房由黑色、不锈钢以及大面积着重使用的玻璃组成。设计使各种物件鲜活起来，使空间更具有打动人的气质。

There are a few key elements in the design that particularly stand out, the first and most impressive of them all is the staircase. This design element is considered the grandest not only because of its creative nature, but because of the impact it had on the design of the house as a whole. Replacing the standard stairwell with a floor to ceiling chandelier by Melograno Blu caused a great challenge for workers in need of constant movement throughout the house as for most of the working time, there was no functional staircase. The result however, was worth the trouble. Standing out as a design masterpiece, the stairway offers a beautiful aesthetic and an interesting twist to what would otherwise be an overlooked structure.
The kitchen is another important design aspect. Considered a "high-tech" kitchen it is kept within Boffi's style of sleek simplicity; appliances and other kitchen utensils are concealed behind clean lines and flushed cabinets while the finishes are a strict combination of black and stainless steel with an accent of a glass table. The design is meant to come alive through the use of its accessories as well as its capacity to welcome personal touches.

STOWE TOWN
斯托小镇

项目名称 斯托小镇 / 设计师 McLean Quinlan / 空间面积 525 m^2
主要建材 橡木、板岩顶板、石墙 / 项目地址 英国英格兰

Project Name Stowe Town / Designer McLean Quinlan / Area about 525 m^2
Materials Oak frame, Slate Roof, Local stone walling / Location England, UK

本项目位于空旷的山区，原本是个谷仓，后来转变为一个住宅区，因此需要全面的大改造。由于之前的转变太小，所以我们决定拆掉原来的建筑，使用拆后可用的砌石重新建造，在其两侧增加一个泳池。
设计的目的是建造一个房屋，使之与整个景观相协调，看起来像是一个传统的农家大院。
业主对我们为之设计的新的度假房屋很满意。房子主要包括一个大的厨房、六个卧室、四个卫浴间、游戏室、杂物间、靴室、健身房和更衣室，还有一个室外泳池。从房子室外和入口处便可知这个房子的实用性必须要强，靴室是设计的重点，它是业主四个孩子和他们朋友进出的重要空间，在乡村，靴子粘上泥土是不可避免的，这个靴室可以使房子的整洁不会受到过多的影响。
这个谷仓的设计和决定创造高度 9m 的单行道使得整个建筑带给人惊喜。结构元素包括绿色的，橡木板顶棚，它使整个大空间有种温暖的感觉，同时一个大的现代壁炉也被设置在这个空间中。
整体来看，内部空间使用的材料颇为简单：科兹窝石地板、橡木地板、横梁、门以及白色喷漆灰泥顶棚。 地下发热装置的能量是由地面热泵产生的，热泵藏在邻近的地面下。

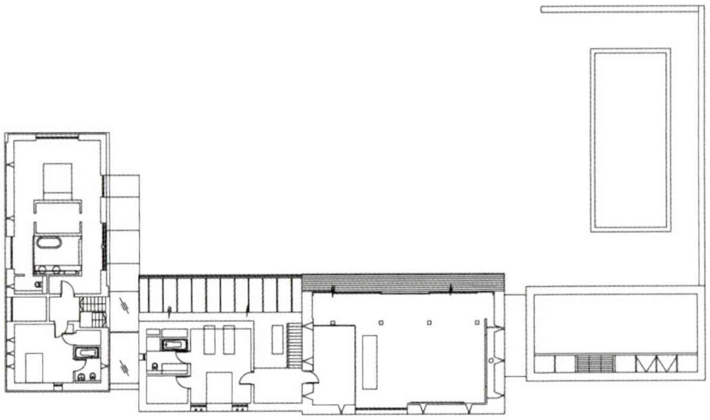

A radical approach was required for this converted barn sited in open countryside. As the previous conversion had been so poor, we decided to demolish the building and through using the salvaged stonework, start again with the addition of two new wings and a swimming pool.

The aim of the design concepts was to create a house that sat naturally in the landscape and appeared to be a traditional agricultural building.

We had worked on the clients' London home and they were keen to have our design ideas for their new weekend/holiday house. The brief included a large kitchen/family room, six bedrooms, four bathrooms, playroom, utility room, boot room, gym and changing rooms and an outdoor pool.

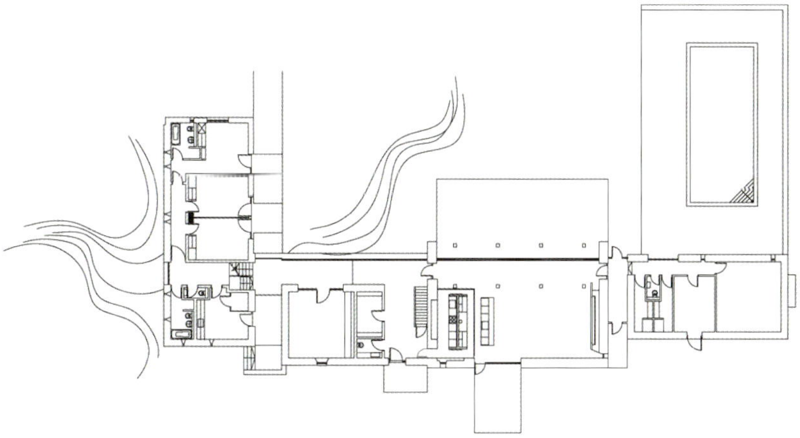

The house needed to be very practical in the way it related to the outside and the entrance and boot room were key to the design, allowing their four children and friends to come and go, while leaving the house relatively unscathed by the inevitable mud and muck of the countryside.

The design of the main barn and the decision to create the 9m high single space created exciting architectural drama. The structural elements were green oak with an oak boarded ceiling to give a sense of warmth to the large space and a large contemporary fireplace was also designed in proportion to the space.

Throughout, the pallet of materials used internally were kept simple: Cotswold stone floors, oak floors, beams, doors and ceilings and white painted plasterwork. The energy to run the under-floor heating is created by ground source heat pumps buried in adjacent land.

GARDEN HOUSE

花园住宅

项目名称 花园住宅 / 设计单位 David Guerra Architecture and Interior / 空间 [illegible] m²
项目地址 巴西米拉斯吉拉斯州贝洛哈里桑塔 / 摄影师 Jomar Bragança

Project Name Garden House / Design Company David Guerra Architecture and [illegible] / Area about 1283 m²
Location Belo Horizonte, Minas Gerais, Brazil / Photographer Jomar Bragança

业主在位于伦敦的某企业任董事一职，这样一处 20 世纪 80 年代风格的住所的设计，是由一位著名的建筑师完成的。业主聘请设计单位做重建和室内的工程设计，所以其中一个主要的目标是使室内外更大化地整合，强调花园这一部分，将更多光线带进室内。由于另一个要点是使住所更现代化，以新方式为住所营造更多新空间，兼容崭新的生活模式。整合空间，拆除墙身，改造门窗，增设起保障作用的覆盖物，替换摆设物件，还原构造、全新的水压和照明设备，此外还有画室、酒窖、美食厨房、洗手间、桑拿房和游泳池。重建工程庞大。

An 80's house. The director of a company, with the head office in London chooses a house in the 80's decade, projected by a big architect. The office was hired to make a reconstruction and interiors project. One of the main wishes was to make the interior/exterior integration bigger, making the garden participate in the day-by-day of the house, bringing more light to the rooms that used to be too dark. Another important point was to make this house more modern, giving new spaces and new ways to use them, compatible with a new lifestyle. Integrate spaces, destroy walls, transform doors, windows and attatch protection covering, substitute products, recuperate structures, new hydraulic, new illumination, an atelier, a wine cellar, gourmet kitchen, bathroom, sauna, deck and pool. The reconstruction was huge.

所有的卫生间、厨房以及服务间墙壁均使用光滑的面料，与木质架构形成对比。只在厨房选用了砂岩，另外还选用了拆卸式木梁与布伦南德陶瓷和花岗岩，与花园的郁郁葱葱一起搭配提升了整体的美感。此住宅的主旨是设计一个与好友相聚的开放式空间，就此，建筑师在室内设计中蓄意营造出完整舒适、乐趣无限、好客热情而又美好的空间氛围。自然的特质，中性而原生态的基调赋予空间舒适感。家具底部离地板的距离恰到好处，扶手椅及亚克力椅使空间显得宽敞、明亮、灵动。住宅的混搭风格，将 20 世纪 50 年代风格、古典风格、民族风与时代感融为一体，是巴西文化的显著体现。厨房、饭厅、卫浴间及服务间采用了白色瓷砖，带出舒适感，同时解决了因采用木架构、木地板和顶棚带来的亮度不足的问题。优质的木质百叶窗使空间更加开阔，壁炉处镶嵌的全新玻璃也为住宅带来良好的光照度，并保持住宅的温馨感。露台被蓄意加大而营造出浪漫氛围的同时又不影响整体的灵动感。旧吧台及食品储藏室与电视间整合，用餐区与厨房亦作开放式整合。迷你衣帽间被改造成衣橱，为卫浴间腾出更多的空间。

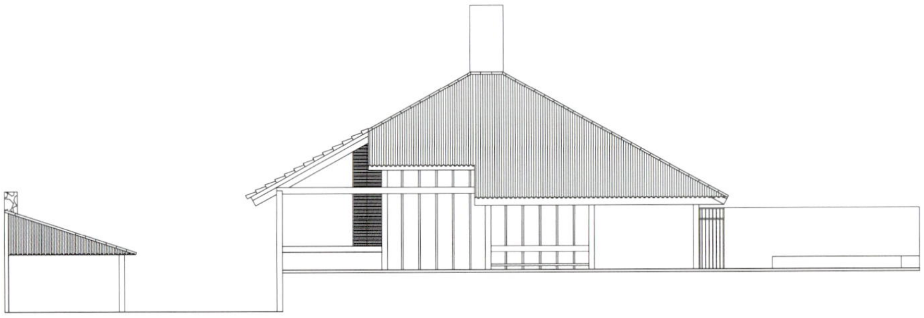

▶

All the bathrooms, kitchen and service clear covering, give an opposite idea of the wood in the structures. Only in the gourmet kitchen the sandstone was chosen, with the use of the demolition wood, ceramic Brennand and granite, enhanced by the proximity from the green in the garden. The idea of the house is to present itself always open for friends, so, the architect found in the interior design the intention to dialogue these integrated and cozy spaces, providing lightness, invitation and meeting, in addition to load with a memory value. Natural characteristic, neutral and raw tones give coziness. Furniture with the bottom distant from the floor, that goes from the stick feet and armchairs, and acrilic chairs have the intetion to make the spaces bigger, light and fluid. The mix of styles, which goes from the 50's, pass by the classics, ethnics, contemporaries, are incremented with brazilian references. The covering of white tile used in the kitchen, dining, bathrooms and service area have the function to give coziness and, at the same time, light up the house, once the structure, floor and ceiling were made of wood. The superior windows in wood shutters gave space to big openings and a new glass coverage appears, in the center of the fireplace, bringing illumination to the house. In the illumination, there was a concern of illuminating well the rooms, keeping this cozy atmosphere. The balconies and the decks got bigger to creat ambience without interfering with the flow. The old bar and pantry are integrated to the TV room, as soon as the integrated dining, without walls, to the kitchen and dining room. The little closets were transformed in wardrobe, expanding the bathrooms.

1001 DISCOS

New York

MOUNTAIN HOUSE
青山住宅

项目名称 青山住宅 / 设计单位 David Guerra Architecture and Interior / 空间面积 745 m^2
项目地址 巴西米拉斯吉拉斯州 / 摄影师 Jomar Bragança

Project Name Mountain House / Design Company David Guerra Architecture and Interior / Area about 745 m^2
Location Nova Lima, Minas Gerais, Brazil / Photographer Jomar Bragança

此案例可以说是周末住所，业主有一位母亲及两个已婚的儿子，还有儿媳妇、孙女。住所所在的场地空旷，周边是生态保护区。我们与业主的想法不谋而合，在此处建一处与自然相结合的住所，可作为冥想和交流的平台。采用大自然的材料以营造温暖、舒适的氛围。
所有私人和公共空间都面朝青山。此处的游廊均被改造成栽满草的庭院，庭院内有带跳板的游泳池、灌水器、水镜和小瀑布。

Weekend refuge for the mother, her two married sons, her daughter-in-law and her grandchildren. It is located in a condominium with big sites, surrounded by an ecological reserve. The idea was to create a house linked to nature which works as a big deck of contemplation and sociability using natural material to make an atmosphere of snugness in a contemporary proposal.
All private and social areas were developed in one single level which faces the mountains. The verandas of these premises are turned to a grassy courtyard where the swimming pool with a deck, a douche, a water mirror and a cascade is located.

为自然、粗犷和现代这几个元素的混搭寻找协调的解决方案是一项挑战，而这些风格的材料正是满足各位业主（母亲和两个儿子）不同品味所必需的。
另一个挑战是统一好钢筋混凝土结构，用于客厅、游廊和卧室，并用可再生的木质材料覆盖在外，外加项目中其他的常规结构系统。选用这些材料的主要目的在于使美感、舒适、安逸得以结合与维系，并在自然与材料间形成反差，像是可再生木料、石料、陶器、竹、瓷砖、纤维、金属、不锈钢、双层釉料、薄层玻璃、玻璃瓦、阳极铝等。住所周边环境恰当地布置，合理利用玻璃和威尼斯木制的百叶窗能够做到使内外空气对流，使空气极其舒适宜人。地板、墙身、顶棚的木料还有大块的木质门可起到吸收噪声的作用，增加了听觉的舒适度。游廊的悬桁屋顶采用钢结构，竹子和聚碳酸酯遮阴，并营造出乡土气息。在家具挑选上，业主并无意随潮流，每一件家具都精心挑选，以反映业主的个人经历，诸如多重性、阅历、价值观。家具涵盖经典的、粗犷的以及现代的风格，均出自巴西设计师之手。所有家具给予整个空间一种明亮感和流动感。

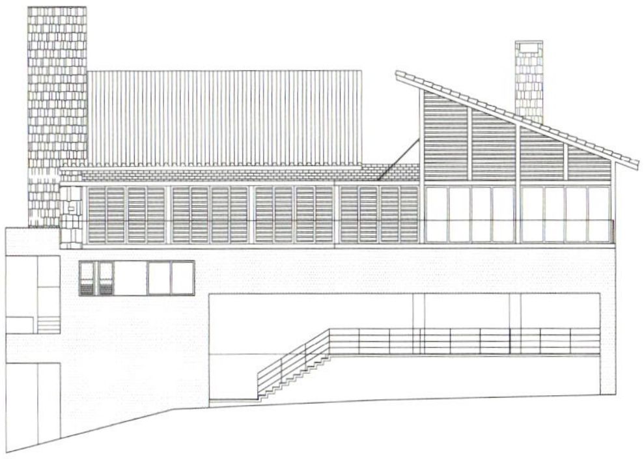

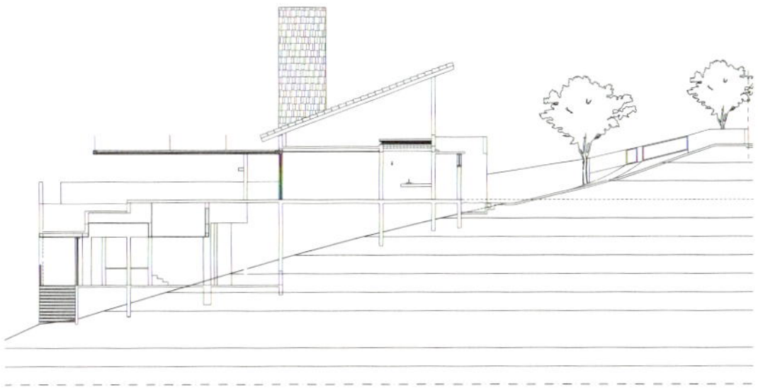

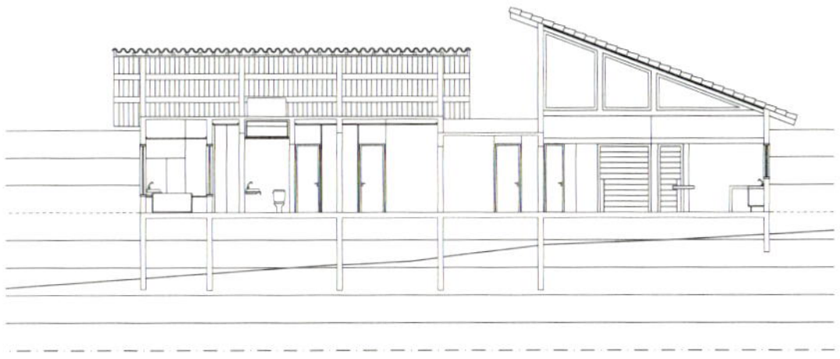

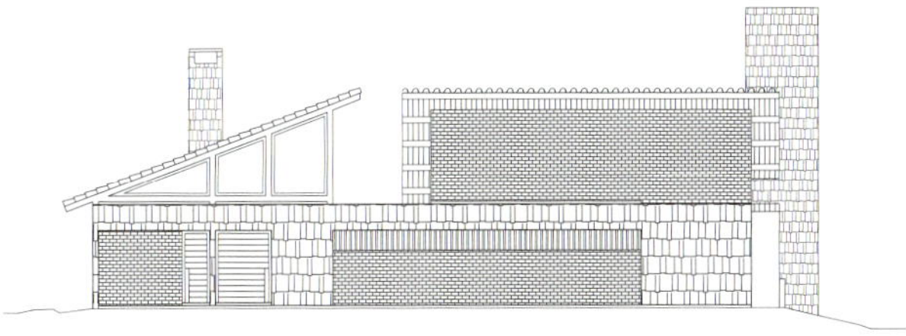

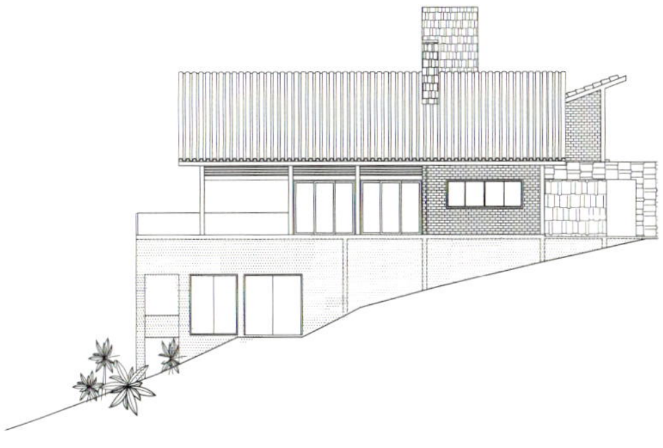

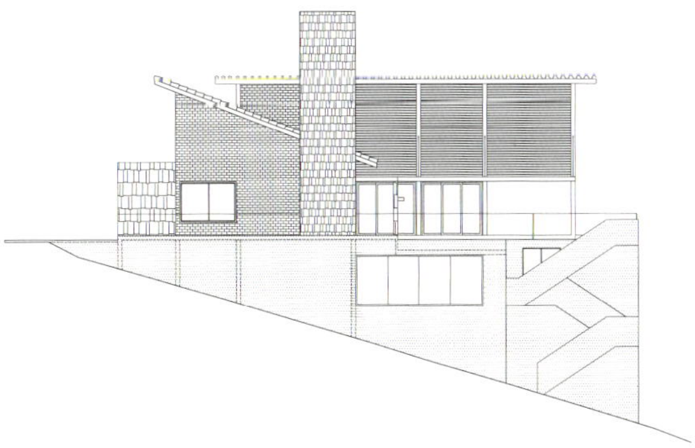

It was challenging to find a harmonious solution for the mixture of natural, rustic and contemporary elements and their materials which were necessary to satisfy the different tastes of the occupants (mother and two sons).
Another challenge was the reconcilement of the steel structure which was used in the living room, the veranda and the bedrooms with its reclaimed wood coating and with the conventional structural system of the rest of the project. The materials were chosen with the aim to combine beauty, snugness and ease of maintenance and to create a contrast between natural and technological materials such as reclaimed wood, stones, earthenware, bamboo, tile, fiber, metal, stainless steel, double glazed and laminated glass, glass tile, anodized aluminum. The appropriate placement of the house within its surroundings, the right usage of glass and the wood Venetian blinds enable cross ventilation which makes the atmosphere extremely comfortable. The use of wood on the floor, the walls and the ceiling as well as the massive wooden doors cause noise absorption which increases the acoustic comfort. The cantilever roofing of the verandas made of steel structure, bamboo and polycarbonate offer shading and create a village atmosphere. There was no intention to follow a fad by the choice of the furniture. Rather each article of furniture was selected carefully to reflect the personal history of the occupants with their multiplicity, their references and values. The furniture includes classic, rustic, contemporary designs and pieces made by Brazilian designers. All the furniture give the entire area a sense of lightness and fluidity.

FOREST HOUSE
森林住宅

森林住宅 / 设计单位 David Guerra Architecture and Interior / 空间面积 500 m²
巴西 / 摄影师 Jomar Bragança

Name Forest House / Design company David Guerra Architecture and Interior
about 500 m² / Location Minas Gerais, Brazil / Photographer Jomar Bragança

此案例是为一对夫妇及其儿子所设计的，位于巴西米拉斯吉拉斯州的贝洛哈里桑塔。住所的总建筑面积为 500 m²，基地面积为 22000 m²。
住所建于当地的树林中，自然景观与现代风格的空间相得益彰。家具均由当代杰出的设计师设计，营造出温暖、舒适的家居氛围。

The project was designed by David Guerra Architecture and Interior in a condominium next to Belo Horizonte, Minas Gerais, Brazil, for a couple with one son. The total construction area of the house is 500 m² and the site area 22000 m².
The house over the native forest combines the nature with the contemporary vision. The furniture includes classics and present designers creating a snug atmosphere.

LE STYLE
PUTMAN

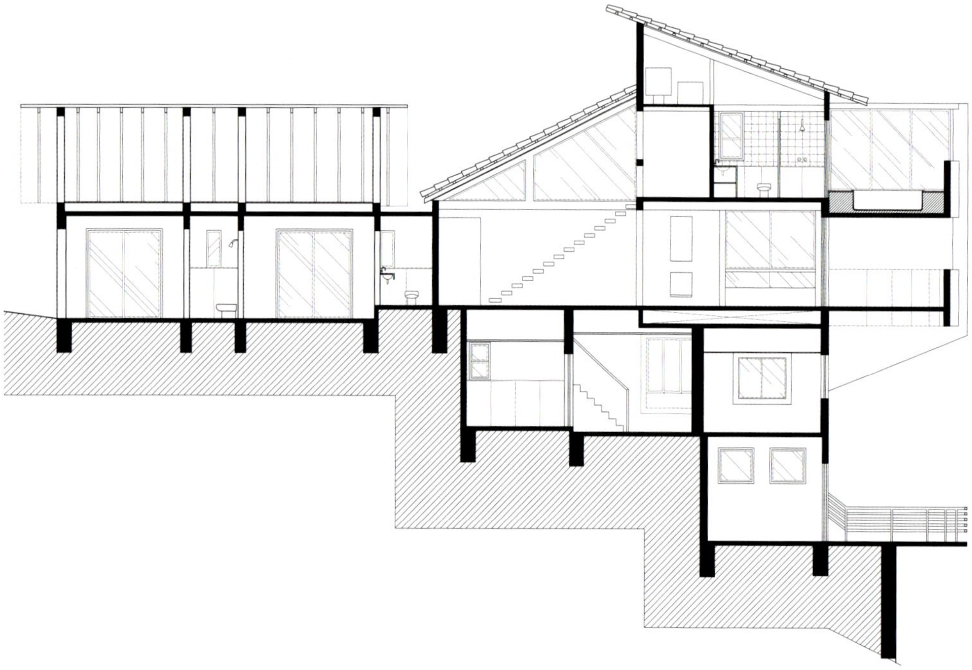

▶
住所的中心是厨房、客厅、饭厅、走廊、泳池、桑拿房、卫生间以及围绕其而建的园林景观。这一切创建出便于交流的大环境。
一楼设有两间卧室和一个衣橱。主人房以及与其配套的办公区、衣帽间、卫生间和阳台均位于二楼，起到了保护隐私的作用。客人公寓间配套的卧室、卫生间和起居室均有独立入口。住所还设有适应性强的酒窖、车库、洗衣房和工人房。
另一个亮点是住所的色彩，明亮的红色与树林的绿色形成极大的反差。地板、顶棚、窗户、门和家具的木质材料加工均出自同一位工匠之手。大面积的玻璃墙使园林景观成为住所的一道风景。简易的单面坡屋顶以及随之改变的水平高度和容积是建筑引人注目的地方。建筑师对住所的室内设计与对家具设计同样的出色。

The living room, the dining room, the verandas, the pool, the sauna, the bathroom and the landscape are developed around the kitchen which is the center of the house. All together work as a big area of sociability.
On the first floor there is also an area with two rooms and a wardrobe. The master suite with its office, closet, bathroom and terrace achieves the privacy on the second floor. The guest "apartment" with its bedroom, bathroom and living room has an independent access. The house also has an acclimated wine cellar, a garage, a laundry and a maid's room.
Other highlight is the color of the house, the vibrant red in contrast with the forest green. The wood of the structure, floor, ceiling, windows, doors and the furniture was made by a carpenter that worked with Zanine Caldas, a famous carpenter in Brazil. The big glass wall allows the landscape turn the scenery of the house. The one slope simple roof is remarkable for the architecture with its levels and its volume change. The architect has to done the house and the interior project as well as the design of the furniture.

图书在版编目（C I P）数据

本色·顶级别墅空间 / 深圳市博远空间文化发展有限公司 编．—武汉：华中科技大学出版社，2011.10
ISBN 978-7-5609-7277-0

Ⅰ．①本… Ⅱ．①深… Ⅲ．①别墅—室内装饰设计—世界—图集 Ⅳ．①TU241-64

中国版本图书馆 CIP 数据核字（2011）第 166846 号

本色·顶级别墅空间 深圳博远空间文化发展有限公司 编

出版发行：华中科技大学出版社（中国·武汉）
地　　址：武汉市武昌珞瑜路 1037 号（邮编：430074）
出 版 人：阮海洪

责任编辑：周　虹　　责任监印：张贵君
责任校对：刘锐帧　　装帧设计：邢晓真

印　　刷：北京利丰雅高长城印刷有限公司
开　　本：965 mm×1270 mm　1/16
印　　张：19.25
字　　数：200 千字
版　　次：2011 年 10 月第 1 版 第 1 次印刷
定　　价：288.00 元　(US $69.99)

华中出版

投稿热线：（010）64155588-8000 hzjztg@163.com
本书若有印装质量问题，请向出版社营销中心调换
全国免费服务热线：400-6679-118　竭诚为您服务